STUDY GUIDE

for

Shannon Donovan

Bridgewater State University

PEARSON

Boston Columbus Indianapolis New York San Francisco Upper Saddle River
Amsterdam Cape Town Dubai London Madrid Milan Munich Paris Montreal Toronto
Delhi Mexico City Sao Paulo Sydney Hong Kong Seoul Singapore Taipei Tokyo

Editor in Chief: Donna Battista
Editorial Project Manager: Mary Kate Murray
Production Project Manager: Alison Eusden

---

www.pearsonhighered.com

PEARSON

ISBN-10: 0-13-387985-2
ISBN-13: 978-0-13-387985-8

Contents

Preface

This study guide is intended to supplement the accompanying textbook, *Principles of Managerial Finance, Brief,* seventh edition. As a supplement, it will review the concepts presented in the text and reinforce the application of these concepts through practice problems and sample exams. The best way to be successful in learning finance is to try working through many different problems that will test your ability to use the concepts you have learned. The field of finance, similar to accounting, is one that requires practice, practice, practice!

This study guide is organized as follows:

Chapter Summary

Learning Goals with a brief description of each.

Chapter Notes with the specific learning goal noted in each section. The learning goals are highlighted on the left side of the page. For example, learning goal 1 is highlighted as follows:

LG 1

Sample Problems and Solutions

Study Tips

Student Notes—a place where you can make your own notes as you work through the chapter

Sample Exam—10–16 True/False questions, 15–26 Multiple Choice questions, and 2–3 Essay questions

Answer Sheet for the Sample Exam

After reading the chapter outline, see if you can work through the sample problems without looking at the solution. Once you have attempted a problem, check your answer with the given solution.

In addition, a sample exam is given to test your knowledge and the application of the presented material. Try taking the exam and then check your answers against the answer sheet. Many of the questions have an explanation for the answer. If you are still unsure of the reason for the answer, review the study guide and/or the textbook.

I wish you great success in learning this material! Your ability to understand the field of finance will greatly advance your career potential regardless of your major and will help you in your personal finance decisions throughout your life. Additionally, the field of finance is an exciting one with many potential career opportunities. In fact, this field is expected to see significant continued growth for employment in the near future.
I wish to thank Larry Gitman and Chad Zutter for producing a truly outstanding managerial finance textbook, my family (Meghan, Connor, and Ryan) for their support, and everyone at Pearson.

Shannon Donovan, D.B.A.
Professor of Finance
Bridgewater State University

Chapter 1
The Role of Managerial Finance

■ Chapter Summary

The *Principles of Managerial Finance, Brief* text used in your finance course is organized around learning goals. Each chapter lists the learning goals, and they are repeated throughout the chapter. Each chapter in this study guide will begin by listing these learning goals along with a brief summary discussion. After reading the chapter outline that follows, see if you can answer the questions raised by the learning goals.

LG 1 *Define finance and the managerial finance function.* Finance is the science and art of managing money. Finance functions involve analyzing the proper allocation of financial assets. The financial analysis applies to business decision making of all types of businesses—private and public, large and small, profit-seeking and not-for-profit. The financial services area encompasses financial product design and the delivery of financial advice. Loan officers, bank managers, and stock brokers are some of the many financial services positions. The managerial finance area encompasses the duties of the financial manager working in a business. Financial analysts, cash managers, and credit analysts perform managerial finance roles.

LG 2 *Describe the legal forms of business organization.* The three most common legal forms of business organization are the sole proprietorship, the partnership, and the corporation. Corporations dominate sole proprietorships and partnerships in terms of receipts. Corporate owners cannot lose more than their investments, while sole proprietors and partners have unlimited liability. The primary emphasis of the text is on corporations.

LG 3 *Describe the goal of the firm, and explain why maximizing the value of the firm is an appropriate goal for a business.* The primary goal of the financial manager is to maximize the wealth of the firm's shareholders. This is not the same thing as maximizing profits. Maximizing profits may result in a short-term management viewpoint and a failure to properly consider risk. To maximize the value of the firm means to run the business in the interest of those who own it, the shareholders. Because shareholders are paid after other stakeholders, it is generally necessary to satisfy the interests of other stakeholders in order to enrich shareholders.

LG 4 *Describe how the managerial finance function is related to economics and accounting.* All areas of responsibility within a firm interact with finance personnel and procedures. The financial manager must understand the economic environment and rely heavily on the economic principle of marginal cost–benefit analysis to make financial decisions. Financial managers use accounting but concentrate on cash flows and decision making.

LG 5 *Identify the primary activities of the financial manager.* The key activities of the financial manager are analyzing and planning, making investment decisions, and making financing decisions.

LG 6 *Describe the nature of the principle–agent relationship between owners and managers of a corporation, and explain how various corporate governance mechanisms attempt to manage agency problems.* This separation of owners and managers of the typical firm is representative of the classic principle–agent relationship, where the shareholders are the principles and managers are the agents. This arrangement works well when the agent makes decisions that are in the principle's best interest but can lead to agency problems when the interests of the principles and agents differ. A firm's corporate governance structure is intended to help ensure that managers act in the best interests of the firm's shareholders, and other stakeholders, and it is usually influenced by both internal and external factors.

■ Chapter Notes

Definition of Finance and the Managerial Finance Function

LG 1 Finance is the science and art of managing money. All individuals and organizations earn or raise money. Finance is concerned with the process, institutions, markets, and instruments involved in the transfer of money among and between individuals, businesses, and governments.

Career opportunities in finance can be broken down into two broad areas:

1. Financial services deal with the design and delivery of advice and financial products to individuals, business, and government. Career opportunities center on banking, financial planning, investments, real estate, and insurance.
2. Managerial finance is concerned with the duties of the financial manager working in a business. The financial manager is responsible for such tasks as budgeting, financial forecasting, cash management, credit administration, investment analysis, and funds procurement.

All managers, regardless of their job descriptions, usually have to provide financial justification for the resources they need to do their job; therefore, understanding the financial aspects of their actions is important to all managers, so they can gain the resources they need to be successful.

Legal Forms of Business Organizations

LG 2 There are three basic legal forms of business organizations, each with its own unique characteristics.

1. A sole proprietorship is owned and operated by one individual. The profits from the sole proprietorship are taxed as the owner's personal income, and the sole proprietor has unlimited liability.
2. Partnerships are formed when two or more individuals enter into a business relationship for the purpose of profits. Like the sole proprietorship, partnerships are taxed as the personal income of each partner, according to that partner's percentage of ownership. Partnerships also have unlimited liability to the creditors of the business.
3. A corporation is a form of business organization that is treated as a separate legal entity from the owners of the business. A corporation can sue and be sued, be party to contracts, and engage in any business function that normally would be performed by individuals. A corporation has limited liability in that the owners' wealth is protected from the corporation's creditors.

The owners of a corporation are its stockholders, whose ownership is evidenced by either common or preferred stock. Stockholders benefit from either an appreciation in the value of their shares or from receipt of dividends. The boards of directors are elected by the stockholders and are responsible for approving strategic goals and plans, setting general policy, guiding corporate affairs, and approving major expenditures. Although only about 20 percent of all U.S. businesses are incorporated, corporations account for roughly 80 percent of total business revenue.

Other organizational forms blend the characteristics of corporations with those of partnerships, such as limited partnerships, S corporations, limited liability companies, and limited liability partnerships.

Goals of the Firm

LG 3

The goal of the financial manager is not to maximize the profits of the firm but to maximize the shareholders' wealth. Profit maximization fails to recognize the timing of returns, cash flows, and risk. The financial manager increases shareholder wealth by taking these issues into account:

1. *Timing of returns*: Because a firm can earn returns on funds it receives, a dollar that is received sooner is much more valuable than a dollar that is received later.
2. *Cash flows available to stockholders*: Financial managers should make decisions that increase the cash flows available to the shareholders. Financial managers can increase cash flows to the stockholders by adopting policies that lead to increased dividends or increased stock prices.
3. *Risk*: The financial manager should examine the tradeoff that exists between risk and return. The higher the risk of a particular project, the higher the return should be to the firm, which leads to increased cash flows for the stockholders.

Firms must also include the interests of stakeholders as well as shareholders. Stakeholders are groups such as employees, customers, suppliers, creditors, owners, and others who have a direct economic link to the firm.

Consider Ethics

Business ethics are standards of conduct or moral judgment that apply to a person engaged in commerce. Financial managers should consider ethical issues when making decisions. An effective ethics program can enhance value.

Relationship of Managerial Finance Function to Economics and Accounting

LG 4

People in all areas of responsibility within the firm must interact with finance personnel and procedures to get their jobs done. Managerial finance has similarities to both economics and accounting.

Finance is closely related to the field of economics in that the financial manager must understand economic theory and be aware of the consequences of varying levels of economic activity and changes in economic policy. The financial manager relies heavily on the economic principle of marginal cost–benefit analysis to make financial decisions. Financial managers use accounting but concentrate on cash flows and decision making.

The fields of finance and accounting are closely related in that both are concerned with the calculation and reporting of information that helps in measuring the performance of a business and assessing its financial position. The main difference between the two fields is that the financial manager must go one step further and make decisions based on the accountant's financial data. Also, accountants compute income using generally accepted procedures, while financial managers concentrate on cash flows.

Key Activities of the Financial Manager

LG 5

The roles of the financial manager can be broken down into three primary activities:

1. Perform financial analysis and planning.
2. Make investment decisions.
3. Make financing decisions.

Principle–Agent Relationships and Corporate Governance

LG 6

The agency problem can be defined as the likelihood that managers may place personal goals ahead of corporate goals. There are two factors that work to contain the agency problem.

1. Market forces constantly monitor the weaknesses of a firm, and any weaknesses that arise due to the agency problem will be corrected by the outside forces.
2. Agency costs, borne by the stockholders of a company, attempt to prevent or minimize the agency problem. There are four main types of agency cost:
   a. Monitoring expenditures
   b. Bonding expenditures
   c. Opportunity costs
   d. Structuring expenditures

Corporate governance refers to the rules, processes, and laws by which companies are operated. Corporate governance tries to reduce or eliminate the principle–agent problem by trying to ensure managers' interests are aligned with shareholders' by approaches such as structuring management compensation to correspond with firm performance. Incentive plans, such as stock options, reduce agency costs by aligning managers' interests more closely to those of stockholders.

Government regulation generally shapes the corporate governance of all firms. The Sarbanes-Oxley Act of 2002 focuses on eliminating many disclosure and conflict of interest problems that were uncovered by the corporate misdeeds of late 1990s. This act did the following:

1. Established an oversight board to monitor the accounting industry
2. Tightened audit regulations and controls
3. Toughened penalties against executives who commit corporate fraud
4. Strengthened accounting disclosure requirements
5. Established board structure and membership guidelines
6. Required instant disclosure of stock sales by corporate executives
7. Increased security regulation authority and budgets for auditors and investigators

■ Sample Problems and Solutions

This section of the study guide will demonstrate how to solve the problems presented in the text. There will be warnings about what mistakes students usually make and detailed explanations about why a particular method is used. Review the problems shown here carefully before working on your homework assignments or the sample test.

Chapter 1 is unique in that it contains few mathematical models. Most of the rest of the text will have many more.

Example 1. Solving for Earning per Share

Profits are commonly expressed in terms of earnings per share (EPS), which represents the amount earned during the period (typically a quarter or a year) on each share of common stock outstanding.

EPS = Total earnings during the period/the number of shares outstanding

If the firm's net income after tax is $100,000 for the year and there are 50,000 shares of common stock in the public's hands, what are the annual earnings per share?

Solution

EPS = $100,000/50,000 = $2

Notes:

1. The total earnings represent the period's earnings that are available for common stockholders. This means that taxes and preferred stock dividends should already be subtracted from the total.
2. The number of shares in the denominator should be only those held by *common stockholders* because preferred stockholders usually receive constant dividends regardless of what the firm earns.

Example 2. Solving Accounting Profit versus Cash Flow

Accounting profits are calculated on an accrual basis where revenue in recognized at the time of sale and expenses recognized when they are incurred. A financial manager uses a cash basis where revenues and expenses are recognized with respect to actual inflows and outflows of cash.

If a firm imports $1,000,000 of goods, which it pays for in cash immediately, and sells all of its goods, during the year for $1,200,000 on account, and has only collected 70 percent of the accounts receivable by year end, what is the accounting profit and what is the financial cash flow?

Solution

Accounting Profit = $1,200,000 – $1,000,000 = $200,000
Financial Cash Flow = (0.7 × $1,200,000) – $1,000,000 = $840,000 – $1,000,000 = –$160,000.

Example 3. Solving for Marginal Cost-Benefit

A primary economic principle used in managerial finance in the marginal cost–benefit analysis. The principle is that financial decisions should be made and actions take when the added benefits exceed the added costs.

A firm believes that if it replaces the current cargo area on its trucks it would produce a total benefit of $750,000 (in today's dollars) over the next 5 years. The current cargo area produces benefits of $500,000 (also in today's dollars) over the same time period. The initial cash investment required for the new cargo area will be $125,000, and the existing cargo area can be sold for $50,000. Apply the marginal cost–benefit analysis to determine what the company should do:

Solution

a) Marginal benefit of the cargo area= Total benefits of new cargo area – Total benefits of the existing cargo area.

   Marginal benefit of the cargo area = $750,000– $500,000 = $250,000

b) Marginal cost of new cargo area = Cost of the new – Sales price of existing cargo area

   Marginal cost of the new cargo area= $125,000-$50,000=$75,000

c) Net benefit= Marginal benefit – Marginal cost = $250,000 – $75,000 = $175,000

d) The firm should recommend the new cargo area for the trucks because the marginal benefit exceeded the marginal cost.

■ Study Tips

In this section of the workbook, you will find notes, tips, and warnings about the chapter to help you complete assignments and prepare for exams. These help notes are the result of teaching finance to thousands of students over many years. Most students have trouble with the same things. This section will try to help you with these tougher parts and to point out areas that justify extra attention because they often show up on exams.

1. Most students learn about the various forms of business in their business law and accounting classes and so are tempted to skim this materia!. Test your knowledge by seeing if you can list the strengths and weaknesses of each form. For example, a primary advantage to the corporate form of business is that the corporation can raise money more easily due to limited liability.

2. The difference between accounting and finance can be summarized quite easily. There are two main factors to remember.
   a. Accountants use an accrual approach to determining income, while financial managers measure cash flows.
   b. The accountants' primary focus is on the collection of data, while the financial managers use the data to make decisions.

3. Know the goal of the financial manager. It is to maximize shareholder wealth. Note that this goal does not say that the financial manager is concerned with bondholder wealth. Shareholder wealth is maximized by maximizing the firm's stock price. This means the financial manager must pay attention to the factors affecting stock price. These include the timing, the risk, and the size of the firm's cash flows.

4. Understand that the actions firms take to prevent managers from maximizing their own wealth and happiness at the expense of the firm are themselves agency costs. For example, the costs of audits are an agency cost because audits would not be necessary if agents were not running the firm.

Student Notes

Sample Exam—Chapter 1

True/False

T F 1. A corporation is considered to be a separate "legal entity" from the owners and has the power of an individual in that it can sue and be sued.

T F 2. The treasurer in medium-to-large-size firms handles the accounting activities, such as tax management, data processing, and financial accounting.

T F 3. The main goal of the financial manager is to increase the firm's overall profits.

T F 4. In a corporation, profits are generally measured in terms of earnings per share.

T F 5. There is a positive correlation between risk and return. The higher the risk of an investment, the higher the required return will be on that investment.

T F 6. The primary role of the accountant in a corporation is to develop and provide financial data that will allow the financial manager to make decisions about the firm's financial position.

T F 7. A firm's earnings are a true representation of cash flows available to the stockholders.

T F 8. Agency costs are fees that are paid by the management of a corporation to compensate any investor who feels that he has suffered a loss due to the agency problem.

T F 9. The accrual method of accounting recognizes revenues at the time of sale and expenses when they are incurred.

T F 10. The cash method of accounting recognizes revenues and expenses only with respect to the actual inflows and outflows of cash.

T F 11. Because of tax benefits, financial managers prefer cash flows later rather than sooner.

T F 12. In general, stockholders wish to incur higher risk in order to ensure higher returns.

T F 13. The stakeholders of a corporation are employees, customers, suppliers, and creditors.

T F 14. The ethical behavior of the financial manager has a substantial impact on the shareholder's wealth maximization concept.

T F 15. A sole proprietor has limited liability, which means that his or her personal wealth is protected from creditors.

T F 16. The capital expenditure manager evaluates and recommends proposed long-term investments.

T F 17. Profit maximization fails to recognize the timing of returns, cash flows, and the risk.

Multiple Choice

1. Which of the following is not one of the primary activities of the financial manager?
   a. Make financing decisions
   b. Prepare financial statements
   c. Make investment decisions
   d. Perform financial analysis

2. A _________ is a form of business organization that is considered an artificial being and has limited liability.
   a. partnership
   b. professional group
   c. corporation
   d. sole proprietorship

3. The main goal of the financial manager is to
   a. maximize profit.
   b. maximize stakeholder wealth.
   c. minimize cost.
   d. maximize shareholder wealth.

4. _________ can be defined as the art and science of managing money.
   a. Economics
   b. Finance
   c. Accounting
   d. Banking

5. Determine earnings per share for Company XYZ in 2014.

   Financial Data of XYZ

Earnings for 2014	$30,000
Outstanding shares of common stock	10,000
Outstanding shares of preferred stock	5,000
Price of stock	$30

   a. $3.00
   b. $2.00
   c. $0.30
   d. $6.00

6. When the risk of an investment is high, the rate of return required by the investor will be
   a. moderate.
   b. low.
   c. high.
   d. equal to the 30-day T-bill.

7. A market force that has in recent years threatened management to perform in the best interest of the shareholder is the _________ in which a firm is acquired by another firm without the support of management.
   a. fidelity bond
   b. market collapse
   c. incentive plan
   d. hostile takeover

8. _________ are groups of individuals, such as employees, customers, suppliers, creditors, and others, who have a direct economic link to a firm.
   a. Shareholders
   b. Stakeholders
   c. Environment
   d. Market forces

9. The _________ has/have the ultimate authority in guiding corporate affairs and in making general policy for the corporation.
   a. chief executive officer
   b. shareholders
   c. stakeholders
   d. board of directors

10. Which of the following is not a positive benefit of an effective ethics program?
    a. It can gain the loyalty, commitment, and respect of the firm's stakeholders.
    b. It can reduce cash flow by changing the perceived risk.
    c. It can reduce potential litigation and judgment costs.
    d. It can maintain a positive corporate imagine and build shareholder confidence.

11. Sue Pena is a financial manager for Target Department Stores, a large chain of department stores operating in the United States. She is currently trying to decide whether to replace one of the firm's inventory systems with a new, more sophisticated one that would both speed processing and handle a larger volume of transactions. The new system would require a cash outlay of $20,000, and the old system could be sold to net $8,000. The total benefits from the new system (measured in today's dollars) would be $15,000. The benefits over a similar time period from the old system (measured in today's dollars) would be $3,000. Applying marginal cost–benefit analysis, what should Sue do?
    a. Sue should not recommend the new system because the firm will experience a net loss of $2,000.
    b. Sue should recommend the new system because the firm will experience a marginal benefit of $12,000.
    c. Sue should recommend the new system because the firm will experience a net gain of $2,000.
    d. Sue should not recommend the new system because the firm will experience a marginal cost of $10,000.

12. Which of the following is a career opportunity in the field of financial service?
    a. Investment advisor
    b. Cash manager
    c. Credit analyst
    d. Corporate financial analyst

13. The most expensive form of business to organize is the
    a. limited partnership.
    b. sole proprietorship.
    c. corporation.
    d. general partnership.

14. The _________ recognizes revenues and expenses only with respect to actual inflows and outflows of cash.
    a. cash method
    b. assignment method
    c. accrual method
    d. revenue realization method

15. The wealth of the shareholders of a corporation is represented by
    a. the price of stock.
    b. earnings per share.
    c. profits.
    d. cash flows.

16. If the managers of a company are not the owners of the company, they are considered
    a. directors.
    b. shareholders.
    c. insiders.
    d. agents.

17. The controller of a company typically is responsible for
    a. financial planning.
    b. strategic planning.
    c. accounting activities.
    d. managing cash.

18. The main economic principle used in managerial finance is
    a. pricing theory.
    b. marginal analysis.
    c. cash flow analysis.
    d. supply and demand.

19. The manager responsible for monitoring and managing the firm's exposure to loss from currency fluctuations is the
    a. foreign exchange manager.
    b. hedge analyst.
    c. currency control manager.
    d. foreign currency analyst.

20. The calendar year for Company XYZ has just ended. During this year, XYZ sold an airplane for \$200,000 that cost them \$100,000 to manufacture. XYZ has not received payment for the sale. Net profit and cash flow for XYZ are
    a. \$100,000 and \$100,000, respectively.
    b. \$200,000 and \$200,000, respectively.
    c. \$100,000 and –\$100,000, respectively.
    d. \$100,000 and –\$200,000, respectively.

21. Pizzas by Mail sold \$100,000 of pizzas in January and \$150,000 in February. Sixty percent of sales are collected the month of the sale, and 40 percent are collected the following month. Cash expenditures are \$80,000 each month. What is the cash flow for February?
    a. \$170,000
    b. \$50,000
    c. \$40,000
    d. \$70,000

22. Investors who are ___________ generally want to avoid risk.
    a. risk-averse
    b. market makers
    c. risk-tolerant
    d. paranoid

23. Which of the following is not included in making investment decisions?
    a. inventory
    b. fixed assets
    c. accounts receivable
    d. notes payable

24. Firms should maximize wealth rather than profits because maximizing profits ignores
    a. the timing of returns.
    b. earnings per share.
    c. cash flows available to stockholders.
    d. risk.

25. While evaluating decision alternatives or potential actions, financial managers must consider which of the following?
    a. only risk
    b. only return
    c. both risk and return
    d. risk, return, and the impact on share price

26. Which of the following is not included in agency costs?
    a. bonding and structuring costs
    b. cost of goods sold
    c. monitoring expenditures
    d. opportunity costs

27. A more recent issue within the business community that is causing major problems is
    a. the privatization of ownership.
    b. short-term versus long-term financial goals of management.
    c. ethical problems.
    d. environmental concerns.

28. Joshua Gilbertstein had the following cash flows in February of this year.

Item	**Amount**
Car payment	$ 415
Dividend received	$ 25
Gas	$ 125
Groceries/food	$ 255
Interest received	$ 44
Net pay received	$3,500
Rent	$ 650
Stock purchase	$1,200

Joshua's net cash flow for the month was
    a. $855.
    b. $924.
    c. $2,124.
    d. $3,569.

Essay

1. Briefly explain the tradeoff between risk and return.

2. Explain why the primary goal of the financial manager should be owner wealth maximization and not profit maximization for the corporation. Give reasons why profit maximization does not work.

3. Explain the principle–agent relationship between owners and managers and how various corporate governance mechanisms attempt to manage this.

4. What is corporate governance?

■ Chapter 1 Answer Sheet

True/False	Multiple Choice	
1. T	1. B	18. B
2. F	2. C	19. A
3. F	3. D	20. C
4. T	4. B	21. B
5. T	5. A	22. A
6. T	6. C	23. D
7. F	7. D	24. D
8. F	8. B	25. D
9. T	9. B	26. B
10. T	10. C	27. D
11. F	11. D	28. B
12. F	12. A	
13. T	13. C	
14. T	14. A	
15. F	15. A	
16. T	16. D	
17. T	17. C	

NP = \$200,000 – \$100,000
CF = –100,000

(0.4 × \$100,000) + (0.6 × \$150,000) – \$80,000
= \$40,000 + \$90,000 – \$80,000 = \$50,000

\$30,000/10,000 = \$3

Benefits with new computer	\$15,000
Less: Benefits with old computer	3,000
(1) Marginal (added) benefits	\$ 12,000
Cost of new computer	\$ 18,000
Less: Proceeds from sale of old computer	8,000
(2) Marginal (added) costs	\$ 10,000
Net benefit [(1) - (2)]	\$ 2,000

Essay

1. Because investors are risk averse, a security must provide a higher level of return to induce investors to accept a higher level of risk. As a result, increasing risk will be associated with increasing return.

2. Profit maximization may result in maximizing profits in each period at the expense of maximizing the stock price. This is because some investments may take a long time to pay off or may have high risk. Ultimately, all that stockholders really care about is the return they earn on holding the stock. Efforts by managers to maximize this return will maximize shareholder wealth. Profit maximization may not work because managers may be tempted to accept riskier projects to increase earnings or to reduce or delay research and development to increase current period earnings.

3. Financial managers are often not only concerned with maximizing shareholder wealth but are also concerned with their personal wealth, job security, and fringe benefits. Corporate governance tries to reduce or eliminate the principle–agent problem by trying to ensure managers' interests are aligned with shareholders' by approaches such as structuring management compensation to correspond with firm performance.

4. Corporate governance refers to the rules, processes, and laws by which companies are operated, controlled, and regulated. It defines the rights and responsibilities of the corporate participants such as the shareholders, board of directors, officers and managers, and other stakeholders, as well as the rules and procedures for making corporate decisions. A well-defined corporate governance structure is intended to benefit all corporate stakeholders by ensuring that the firm is run in a lawful and ethical fashion, in accordance with best practices, and subject to all corporate regulations. A firm's corporate governance is influenced by both internal factors such as the shareholders, board of directors, and officers as well as external forces such as clients, creditors, suppliers, competitors, and government regulations

Chapter 2
The Financial Market Environment

■ Chapter Summary

In this chapter you will learn about finance institutions and finance markets. You will learn the difference between primary and secondary markets, money and capital markets, and dealer and broker markets. You will learn about how the government regulates these markets and about how a finance crisis can occur. There will also be a discussion on business taxes and their importance in financial decisions. By the end of this chapter, you will have a much better understanding of the financial market environment.

LG 1 *Understand the role that financial institutions play in managerial finance.* Financial institutions bring net suppliers of funds and net demanders of funds together to help translate the savings of individuals, businesses, and governments into loans and other types of investments. The net suppliers of funds are generally individuals or households who save more money than they borrow. Businesses and governments are generally net demanders of funds, meaning that they borrow more money than they save.

LG 2 *Contrast the functions of financial institutions and financial markets.* Both financial institutions and financial markets help businesses raise the money they need to fund new investments for growth. Financial institutions collect the savings of individuals and channel those funds to borrowers, such as businesses and governments. Financial markets provide a forum in which savers and borrowers can transact business directly. Businesses and governments issue debt and equity securities directly to the public in the primary market. Subsequent trading of these securities between investors occurs in the secondary market.

LG 3 *Describe the differences between the capital markets and the money markets.* In the money market, savers who want a safe, temporary place to deposit funds where they can earn interest interact with borrowers who have a short-term need for funds, generally a year or less. In contrast, the capital market is the forum in which savers and borrowers interact on a long-term basis (more than a year).

LG 4 *Explain the root causes of the 2008 financial crisis and recession.* The financial crisis was caused by several factors related to investments in real estate. Financial institutions lowered their standards for lending to prospective homeowners, and institutions also invested heavily in mortgage-backed securities. When home prices fell and mortgage delinquencies rose, the value of the mortgage-backed securities held by banks plummeted, causing some banks to fail and many others to restrict the flow of credit to business. That in turn contributed to a severe recession in the United States and abroad.

LG 5 *Understand the major regulations and regulatory bodies that affect financial institutions and markets.* The Glass-Steagall Act created the FDIC and imposed a separation between commercial and investment banks. The act was designed to limit the risks banks could take and to protect depositors. Recently, the Gramm-Leach-Bliley Act essentially repealed the elements of Glass-Steagall pertaining to the separation of commercial and investment banks. After the recent financial crisis, much debate has occurred regarding the proper regulation of large financial institutions.

The Securities Act of 1933 and the Securities Exchange Act of 1934 are the major pieces of legislation shaping the regulation of financial markets. The 1933 act focuses on regulating the sale of securities in the primary market, whereas the 1934 act deals with regulations governing transactions in the secondary market. The 1934 act also created the Securities and Exchange Commission, the primary body responsible for enforcing federal securities laws.

LG 6 *Discuss business taxes and their importance in financial decisions.* Corporate income is subject to corporate taxes. Corporate tax rates apply to both ordinary income (after deduction of allowable expenses) and capital gains (when an asset is sold for more than its initial purchase price). The average tax rate paid by a corporation ranges from 15 to 35 percent. Corporate taxpayers can reduce their taxes through certain provisions in the tax code: dividend income exclusions and tax-deductible expenses.

■ Chapter Notes

The Role of Financial Institutions in Managerial Finance

LG 1 Firms can obtain funds from external sources in three ways:

1. Financial institutions, which serve as intermediaries, accepting savings and transferring them to those who need funds
2. Financial markets, which are organized forums in which the suppliers and demanders of various types of funds can make transactions
3. Private placement is the sale of a new security directly to an investor or group of investors

The net suppliers of funds are generally individuals or households who save more money than they borrow. Businesses and governments are generally net demanders of funds, meaning that they borrow more money than they save.

The major financial institutions in the U.S. economy are commercial banks, savings and loans, credit unions, savings banks, insurance companies, mutual funds, and pension funds. Investment banks are institutions that assist companies in raising capital, advise firms on major transactions, and engage in trading and market making activities. The shadow banking system, which grew in the 1990s, is when a group of institutions engage in lending activities, like traditional banks, but do not accept deposits and, therefore, are not subject to the same regulations as traditional banks.

Contrast Functions of Financial Institutions and Financial Markets

LG 2

Both financial institutions and financial markets help businesses raise the money they need to fund new investments for growth. However, financial institutions collect the savings of individuals and channel those funds to borrowers such as businesses and governments, whereas financial markets provide a forum in which savers and borrowers can transact business directly. Businesses and governments issue debt and equity securities directly to the public in the primary market. Subsequent trading of these securities between investors occurs in the secondary market.

Differences between the Capital Markets and the Money Markets

LG 3

Two key financial markets are the money market and the capital market.

In the money market, the investors are savers who want a safe, temporary place to deposit funds where they can earn interest and interact with borrowers who have a short-term need for funds, generally a year or less. Marketable securities including Treasury bills, commercial paper, and other instruments are the primary securities traded in the money market. The Eurocurrency market is the international equivalent of the domestic money market.

In contrast, the capital market is the forum in which savers and borrowers interact on a long-term basis, more than a year. Firms issue either debt (bonds) or equity (common and preferred stock) securities in the capital market. The Eurobond market typically issues bonds denominated in dollars and sells them to investors located outside the United States. The international equity market allows corporations to sell blocks of shares to investors in a number of different countries simultaneously.

Once issued, these securities trade on secondary markets that are either broker markets or dealer markets. In a broker market, for the trade to take place, the buyer and seller are brought together, and the securities change hands on the floor of an exchange. In the dealer market, the buyer and seller are never brought together directly; instead, market makers execute the orders through an electronic trading platform.

An important function of the capital market is to determine the underlying value of the securities issued by businesses. In an efficient market, the price of a security is an unbiased estimate of its true value. An efficient market allocates funds to their most productive uses as a result of competition among wealth-maximizing investors, and thus, as investors learn new information, it is reflected quickly in stock prices.

Root Causes of the 2008 Financial Crisis and Recession

LG 4

The 2008 financial crisis was caused by several factors related to investments in real estate and spread to other industries, which resulted in a severe global recession.

Securitization is the process of pooling mortgages or other types of loans and then selling claims or securities against that pool in the secondary market. These securities are called mortgage-backed securities and can be purchased by virtually any investor. The basic risk of these securities is that homeowners may not be able or may choose not to repay their loans. When home prices are rising and borrowers are having difficulty making payments on their mortgages, lenders allow borrowers to tap this built-up equity and refinance. When home prices are falling, there is no built-up equity, and in fact the price of the home may fall below the value of the mortgage, causing borrowers to walk away from their homes.

Securitization made it easier for banks to lend money because they could pass the risk on to other investors. Soon financial institutions lowered their standards for lending to prospective homeowners, and institutions also invested heavily in mortgage-backed securities. When home prices fell and mortgage delinquencies rose, the value of the mortgage-backed securities held by banks plummeted, causing some banks to fail and many others to restrict the flow of credit to business. That in turn contributed to a severe recession in the United States and abroad.

Major Regulations and Regulatory Bodies That Affect Institutions and Markets

LG 5

Governments regulate financial institutions and markets to avoid financial crisis. The biggest benefit of government regulation is the resulting trust and confidence in the financial institutions and markets derived by society.

In 1933, during the Great Depression, the Glass-Steagall Act created the FDIC and imposed a separation between commercial and investment banks. The act was designed to limit the risks that banks could take and to protect depositors. Recently, the Gramm-Leach-Bliley Act essentially repealed the elements of Glass-Steagall pertaining to the separation of commercial and investment banks. After the recent financial crisis, much debate has occurred regarding the proper regulation of large financial institutions. The Dodd-Frank Act was passed in 2010 and contained a host of new regulatory requirements, the effects of which are yet to be determined.

The Securities Act of 1933 and the Securities Exchange Act of 1934 are the major pieces of legislation shaping the regulation of financial markets. The 1933 act focuses on regulating the sale of securities in the primary market, whereas the 1934 act deals with regulations governing transactions in the secondary market. The 1934 act also created the Securities and Exchange Commission, the primary body responsible for enforcing federal securities laws.

Business Taxes and Their Importance in Financial Decisions

LG 6

Corporate income is subject to corporate taxes. Corporate tax rates apply to both ordinary income (after deduction of allowable expenses) and capital gains. Ordinary income is income earned through the sale of goods or services. The more is earned, the greater the percentage of tax owed. The average tax rate paid by corporations ranges from 15 to 35 percent. Corporate taxpayers can reduce their taxes through certain provisions in the tax code: dividend income exclusions and tax-deductible expenses.

The average tax is the amount of tax paid divided by total income. The marginal tax rate is the percentage of tax owed on the next dollar earned. Marginal rates usually exceed average rates.

Capital gains are income earned due to the sale of an asset for more than its initial purchase price. Currently, capital gains are added to ordinary income and taxed at the regular corporate rate, with the maximum rate of 39 percent. To simplify tax calculations, a fixed rate of 40 percent will be used throughout the remaining chapters in this text regardless of whether it is ordinary or capital gain income.

■ Sample Problems and Solutions

Example 1. Income tax calculation

A firm has $1,000,000 in ordinary and capital gains income. What will this firm owe in taxes according to the figures in Table 2.1 of the text? Calculate the average tax rate and the marginal tax rate on the basis of your findings.

Solution

Total taxes due = $113,900 + [0.34 × ($1,000,000 – $335,000)]

Total taxes due = $113,900 + (0.34 × $665,000)

Total taxes due = $113,900 + $226,100 = $340,000

Average tax rate = $340,000/$1,000,000 = 0.34 = 34%

Marginal tax rate = 34%

Example 2. Interest versus dividend expense

A firm expects earnings before interest and taxes to be $60,000 for this period. Assuming an ordinary tax rate of 40 percent, compute the firm's earnings after taxes and earnings available for common stockholders (earnings after taxes and preferred stock dividends, if any) under the following conditions:

a. The firm pays $20,000 in interest.

b. The firm pays $20,000 in preferred stock dividends.

Solution

a. Taxes of 40 percent will be paid on earnings before interest and taxes of $60,000 less the interest of $20,000. Therefore the taxes will be $16,000 [= (60,000 – 20,000) × 0.4]. So the earnings after taxes and the earnings available for common shareholders is $24,000 [= 60,000 – 20,000 – 16,000].

b. Taxes of 40 percent will be paid on the earnings before interest and taxes of $60,000. Therefore the taxes will be $24,000 [= 60,000 × 0.4]. So the earnings available after taxes is $36,000 [60,000 – 24,000]. The earnings available to common shareholders are the earnings available after taxes less the preferred stock dividends, $16,000 [= 36,000 – 20,000].

   We can see from the above that because interest is a tax deductible expense, the earnings available to shareholders is greater when a firm pays interest rather than preferred dividends.

■ Study Tips

1. This chapter contains many terms and definitions. It is important to learn these terms to fully understand the financial environment in which the rest of the book takes places. Make sure you know the differences between financial institutions and finance markets, between the primary and secondary market, and between the money and capital market.

2. Understanding when and why different regulations were written will help you understand the current financial environment.

3. Although it is important to understand how corporate taxes are calculated with a table as explained in this chapter, remember that throughout the rest of the book we assume a 40 percent marginal tax rate.

■ Student Notes

■ Sample Exam—Chapter 2

True/False

T F 1. Individuals or households are generally the net suppliers of funds and businesses, and governments are generally the net demanders of funds.

T F 2. Businesses and governments issue debt and equity directly to the public in the secondary market.

T F 3. Financial markets are organized forums in which the suppliers and demanders of various types of funds can make transactions.

T F 4. Capital markets are for investors who want a safe, temporary place to deposit funds where they can earn interest and interact with borrowers who have a short-term need for funds.

T F 5. Money markets are markets for long-term funds, such as bonds and equity.

T F 6. An efficient market allocates funds to their most productive uses as a result of competition among wealth-maximizing investors.

T F 7. Securitization is the process of pooling mortgages or other types of loans and selling the claims or securities against that pool in the secondary market.

T F 8. Securitization made it harder for banks to lend money because they could pass the risk on to other investors.

T F 9. The Glass-Steagall Act was imposed to encourage commercial and investment banks to combine and work together.

T F 10. The Securities Act of 1933 focuses on regulating the sale of securities in the primary market, whereas the 1934 act deals with regulations governing the transactions in the secondary market.

T F 11. Corporate tax rates apply only to ordinary income, not capital gains.

T F 12. Capital gains are income earned due to an increase in the value of the asset.

T F 13. The marginal tax is the amount of the tax paid divided by total income.

T F 14. The average tax rate represents the rate at which the next dollar of income is taxed.

T F 15. The biggest benefit of government regulation is the resulting trust and confidence in the financial institutions and markets derived by society.

Multiple Choice

1. Successful firms can obtain funds from which of the following external sources?
   a. Financial institutions
   b. Financial markets
   c. Private placement
   d. All of the above

2. ________ serve as intermediaries, channeling the savings of individuals, businesses, and governments into loans and investments.
   a. Financial markets
   b. Financial institutions
   c. Securities Exchanges
   d. OTC Markets

3. ____________ are generally the net suppliers of funds for financial institutions.
   a. Individuals
   b. Governments
   c. Businesses
   d. All of the above

4. ________ are generally the net demanders of funds for financial institutions.
   a. Individuals and governments
   b. Individuals and businesses
   c. Businesses and governments
   d. Governments

5. ______________ provide savers with a secure place to invest funds and offer both individuals and companies loans to finance investments.
   a. Investment banks
   b. Securities exchanges
   c. Mutual funds
   d. Commercial banks

6. ____________ assist companies in raising capital, advise firms on major transactions such as mergers or financial restructuring, and engage in trading and market-making activities.
   a. Investment banks
   b. Securities exchanges
   c. Mutual funds
   d. Commercial banks

7. The Glass-Steagall Act
   a. allowed commercial and investment banks to engage in the same activities.
   b. created the Securities Exchange Commission.
   c. created the Federal Deposit Insurance program and separated the activities of commercial and investment banks.
   d. was intended to regulate the activities in the primary market.

8. __________ are forums in which suppliers of funds and demanders of funds can transact business directly.
   a. Shadow banking systems
   b. Financial markets
   c. Commercial banks
   d. Financial institutions

9. The sale of a new security directly to an investor or group of investors is called
   a. the secondary market.
   b. the primary market.
   c. the capital market.
   d. a private placement.

10. The ___________ market is where securities are initially issued, and the __________ market is where pre-owned securities (not new issues) are traded.
    a. primary; secondary
    b. secondary; primary
    c. money; capital
    d. primary; money

11. The _______ market is created by a financial relationship between suppliers and demanders of short-term funds, whereas the __________ market enables suppliers and demanders of long-term funds to make transactions.
    a. capital; money
    b. money; capital
    c. primary; secondary
    d. secondary; primary

12. The key capital market securities are
    a. negotiable certificates of deposit
    b. commercial paper
    c. U.S. treasuries
    d. bonds and common and preferred stock

13. The key money market securities are
    a. corporate and municipal bonds .
    b. common stocks and preferred stocks.
    c. treasury bills and negotiable certificates of deposit.
    d. Eurobonds.

14. In a __________ market, the buyer and seller are brought together to trade securities in an organization called __________________.
    a. dealer; securities market
    b. broker; over-the-counter market
    c. broker; securities market
    d. dealer; over-the-counter market

15. In a ___________ market, the buyer and seller are not brought together directly but instead have their orders executed on the ______________.
    a. dealer; securities market
    b. broker; over-the-counter market
    c. broker; securities market
    d. dealer; over-the-counter market

16. The _____ price is the highest price offered to purchase a security, and the ____ price is the lowest price at which a security is offered for sale.
    a. market; auction
    b. bid; ask
    c. ask; bid
    d. efficient; market

17. An efficient market is one where prices of stocks
    a. move up and down widely without any apparent reason.
    b. remain steady for long periods of time.
    c. are steadily decreasing for long periods of time.
    d. incorporate new information quickly and adjust appropriately to their true value.

18. An implication of the Efficient Market Hypothesis is that it is very hard for an actively managed mutual fund to earn above average returns. This is true for all EXCEPT which of the following reasons?
    a. Predictable information is already incorporated into stock prices.
    b. New information is by definition unpredictable, thus hard to incorporate into stock prices.
    c. Actively managed mutual funds typically charge fees of about 1.5 percent.
    d. Index funds make no attempt to analyze stocks.

19. The process of pooling mortgages or other types of loans and then selling claims or securities against that pool in the secondary market is called
    a. refinancing.
    b. securitization.
    c. private placement.
    d. pooling.

20. The primary risk of mortgage-backed securities is that
    a. prices of housing will increase.
    b. interest rates will go down.
    c. the government will not be able to meet their guarantees on the cash flows.
    d. homeowners may not be able or choose not to repay their loans.

21. A crisis in the financial sector often spills over into other industries because when financial institutions ________ borrowing, activity in most other industries also ________.
    a. increase; slows down
    b. contract; slows down
    c. increase; increases
    d. contract; increases

22. The Federal Deposit Insurance Corporation (FDIC) guarantees individuals will not lose
    a. any money held at a bank that fails.
    b. any money, up to a specified amount, held at a bank that fails.
    c. any money held at any type of financial institution that fails.
    d. any money, up to a specified amount, held at type of financial institution that fails.

23. The Gramm-Leach-Biley Act
    a. allows business combinations between commercial banks, investment banks, and insurance companies.
    b. does not allow business combinations between commercial banks, investment banks, and insurance companies.
    c. allows business combinations between commercial banks and investment banks, but not insurance companies.
    d. was signed right after the Great Depression because of the financial crisis.

24. ________________ regulates the secondary market and created the Securities Exchange Commission (SEC).
    a. The Securities Act of 1933
    b. The Gramm-Leach-Biley Act
    c. The Securities Exchange Act of 1934
    d. The Glass-Steagall Act

24. Use the tax rate schedule in Table 2.1 to find the tax on a company that has before-tax earnings of $300,000.
    a. $117,000
    b. $78,000
    c. $139,250
    d. $100,250

26. Use the tax rate schedule on Table 2.1 to find the marginal tax rate on a company who has before-tax earnings of $300,000.

    a. 15 percent
    b. 39 percent
    c. 34 percent
    d. 25 percent

27. Use the tax rate schedule on Table 2.1 to find the average tax rate on a company who has before-tax earnings of $300,000.

    a. 33 percent
    b. 46 percent
    c. 26 percent
    d. 39 percent

28. If two firms have the same earnings before interest and taxes and Firm A pays $10,000 in interest expenses but no preferred dividend and Firm B pays no interest expense but $10,000 in preferred dividends, which firm has the most earnings available for common shareholders?

    a. Firm A
    b. Firm B
    c. They will be equal
    d. Need more information to determine

Essay

1. Describe the differences between the money market and the capital market.

2. Contrast broker and dealer markets.

3. Explain the root causes of the recent financial crisis and recession.

■ Chapter 2 Answer Sheet

True/False

1. T
2. F
3. T
4. F
5. F
6. T
7. T
8. F
9. F
10. T
11. F
12. T
13. F
14. F
15. T

Multiple Choice

1. D
2. B
3. A
4. C
5. B
6. A
7. C
8. B
9. D
10. A
11. B
12. D
13. C
14. C
15. D
16. B
17. D
18. D
19. B
20. D
21. B
22. B
23. A
24. C — $22,250 + 0.39 × ($300,000 – $100,000)
25. D
26. B
27. A — ($22,250 + 0.39 × ($300,000 – $100,000))/$300,000
28. A

Essay

1. In the money market, savers who want a temporary place to deposit funds where they can earn interest interact with borrowers who have a short-term need for funds, generally a year or less. In contrast, the capital market is the forum in which savers and borrowers interact on a long-term basis, more than a year.

2. The key difference between broker and dealer markets is the way in which trades are executed. The broker market consists of national and regional securities exchanges. With the assistance of brokers, the buyer and seller are simultaneously brought together on the floor of this exchange. In contrast, in the dealer market there are essentially two trades in every transaction, with the buyer purchasing shares from a dealer and the seller selling shares to a dealer. Broker markets include the New York Stock Exchange, while the dealer market includes Nasdaq and the over-the-counter market. Because investors are risk averse, a security must provide a higher level of return to induce investors to accept a higher level of risk. As a result, increasing risk will be associated with increasing return.

3. The financial crisis was caused by several factors related to investments in real estate. Financial institutions lowered their standards for lending to prospective homeowners, and institutions also invested heavily in mortgage-backed securities. When home prices fell and mortgage delinquencies rose, the value of the mortgage-backed securities held by banks plummeted, causing some banks to fail and many others to restrict the flow of credit to business. That in turn contributed to a severe recession both in the United States and abroad.

Chapter 3
Financial Statements and Ratio Analysis

■ Chapter Summary

This chapter demonstrates the use of ratios to help examine the health of a firm. Ratio analysis is useful for identifying both problem areas and areas of strength for a firm. One ratio by itself is usually not very useful. It is when ratios are compared—both across time and to those of similar firms—that the benefits are realized. By the end of this chapter, you will find yourself much more familiar with balance sheets and income statements.

LG 1 *Review the contents of the stockholders' report and the procedures for consolidating international financial statements.* The stockholders' report contains the letter to stockholders, four key financial statements, and notes to the financial statements.

LG 2 *Understand who uses financial ratios, and how.* Both insiders and outsiders use financial ratio analysis to compare a firm's performance and status to that of other firms or to itself over time. Financial statement analysis takes two forms: cross-sectional analysis, where firms are compared to other similar firms, and time-series analysis, where firms are compared to themselves at different points in time.

LG 3 *Use ratios to analyze a firm's liquidity and activity.* It is best to perform ratio analysis by grouping ratios together that examine a common issue. For example, activity ratios measure the speed with which various accounts are converted into sales or cash. Liquidity ratios measure the firm's ability to pay its bills by examining the net working capital, current ratio, or quick ratio.

LG 4 *Discuss the relationship between debt and financial leverage and the ratios used to analyze a firm's debt.* The debt ratio and the debt/equity ratio measure indebtedness. Coverage ratios, such as times interest earned and fixed payment coverage, measure the ability to service fixed contractual requirements such as interest, principal, or sinking-fund payments.

LG 5 *Use ratios to analyze a firm's profitability and its market value.* The common-size statement can be used to examine the gross profit margin, the operating profit margin, and the net profit margin. Other measures of profitability include the return on assets, the return on equity, and earnings per share. The market ratios include the price/earnings ratio and the market/book ratio.

LG 6 *Use a summary of financial ratios and the DuPont system of analysis to perform a complete ratio analysis.* A summary of all ratios can be used to perform a complete ratio analysis using cross-sectional and time-series analysis. The DuPont system provides a framework for dissecting the firm's overall financial statements and assessing its condition. The focal point of the DuPont system is the return on total assets. This is explained by the net profit margin and total asset turnover. If the ROA indicates a problem, subordinate ratios can be examined to identify the source of the problem.

■ Chapter Notes

Stockholders' Report and Procedures for Consolidating International Financial Statements

LG 1

Corporations are required to produce annual stockholders' reports that are sent to all shareholders. It will usually begin with a letter from management to stockholders that summarizes the state of the firm and management's views as to its future. This letter is then followed by the four basic statements:

1. *Income statement*: A financial summary of the firm's operating results during a specified period.
2. *Balance sheet*: A summary of the firm's financial position at a given time.
3. *Statement of Shareholders' equity or the abbreviated statement of retained earnings:* Reconciles the net income earned during a given year, and any cash dividends paid, with the change in retained earnings between the start and the end of that year.
4. *Statement of cash flows*: A summary of the cash flows over the period of concern.

The stockholders' report will conclude with notes to the financial statement. These provide detailed information on the accounting policies, procedures, calculations, and transactions underlying entries in the financial statements.

The guidelines used to prepare and maintain financial records and reports are known as generally accepted accounting principles (GAAP). These accounting practices and procedures are authorized by the accounting profession's rule-setting body, the Financial Accounting Standard Board (FASB). In addition, auditors of public corporations are overseen by Public Company Accounting Oversight Board (PCAOB) established by the Sarbanes-Oxley Act of 2002.

U.S.-based companies must consolidate their foreign and domestic financial statements by translating their foreign-currency denominated assets and liabilities into dollars using the current rate (translation) method.

Use of Financial Ratios

LG 2

Ratio analysis is used to compare a firm's performance and status with that of other firms or to itself over time.

There are two types of ratio comparisons: cross-sectional and time-series.

1. Cross-sectional analysis compares different firms' financial ratios at the same point in time. It involves comparing the firm's ratios to those of an industry leader or to the industry averages. Benchmarking is often used in cross-sectional analysis.
2. Time-series analysis is the evaluation of a firm's performance over time. Time-series allows the firm to compare its current performance to past performance.

Cautions about using ratio analysis.

1. Ratios with large deviations from the norm merely indicate symptoms of the possibility of a problem, indicating further investigation is needed.
2. Do not use a single ratio to judge the overall performance of the firm.
3. The financial statements being compared should be dated at the same point of time during the year.
4. Audited financial data should be used to ensure relevant financial information.
5. The financial data being compared in the ratio analysis should be developed in the same manner.
6. When performing time-series analysis, inflation should always be taken into account.

Liquidity Ratios

LG 3

The liquidity of a corporation measures its ability to satisfy its short-term obligations. In other words, liquidity measures the ease with which it can pay its bills. The three basic measures of liquidity are net working capital, the current ratio, and the quick (acid-test) ratio.

Net working capital is used to measure the firm's overall liquidity.

$$\text{Net Working Capital} = \text{Current Assets} - \text{Current Liabilities}$$

The current ratio measures the firm's ability to meet its short-term obligations.

$$\text{Current Ratio} = \frac{\text{Current Assets}}{\text{Current Liabilities}}$$

> More Liquidity

The quick (acid-test) ratio is similar to the current ratio except that the quick ratio excludes inventory, which is generally the least liquid current asset. Therefore, this ratio is preferred over the current ratio when inventory cannot be easily converted into cash.

$$\text{Quick Ratio} = \frac{\text{Current Assets} - \text{Inv}}{\text{Current Liabilities}}$$

> More Liquidity

Activity Ratios

Activity ratios are used to measure the speed with which various accounts are converted into sales or cash.

Inventory turnover measures the liquidity of a firm's inventory.

$$\text{Inventory Turnover} = \frac{\text{Cost of Goods Sold}}{\text{Inventory}}$$

The average collection period is useful in evaluation of credit and collection policies.***

$$\text{Average Collection Period} = \frac{\text{Accounts Receivable}}{\text{Average Sales Per Day}}$$

The average payment period measures the average amount of time needed to pay accounts payable.

$$\text{Average Payment Period} = \frac{\text{Accounts Payable}}{\text{Average Purchases Per Day}}$$

The total asset turnover indicates the efficiency with which the firm uses all its assets to generate sales.

$$\text{Total Asset Turnover} = \frac{\text{Sales}}{\text{Total Assets}}$$

*The average collection period is meaningful in relation to the firm's credit terms and the average payment period is meaningful in relation to the credit terms extended to the firm.

Analyzing Debt

LG 4

Creditors' claims must be satisfied before earnings are distributed to the shareholders, so it is in the best interest of the present and prospective shareholders to pay close attention to the indebtedness of a corporation.

Financial leverage is a term is used to describe the magnification of risk and return introduced through the use of fixed cost financing, such as debt and preferred stock. The more debt a firm uses, the greater its financial leverage, which increases its risk and expected return.

The debt ratio measures the proportion of total assets financed by the firm's creditors. The higher this ratio, the more financial leverage the firm has.

$$\text{Debt Ratio} = \frac{\text{Total Liabilities}}{\text{Total Assets}}$$

The debt-to-equity ratio measures the relative proportion of total liabilities to common stock equity used to finance the firm's assets.

$$\text{Debt-to-Equity} = \text{Total liabilities} \div \text{Common stock equity}$$

The times-interest-earned ratio measures the firm's ability to make interest payments.

$$\text{Times Interest Earned} = \frac{\text{Earnings Before Interest and Taxes}}{\text{Interest}}$$

The fixed payment coverage ratio measures the firm's ability to meet all fixed payment obligations, such as loan interest and principal, lease payments, and preferred stock dividends.

$$\text{Fixed Payment Coverage} = \frac{\text{Earnings Before Interest and Taxes} + \text{Lease Payments}}{\text{Int} + \text{Lease} + \{(\text{Prin} + \text{Pref Stock Div}) \times [1/(1-\text{T})]\}}$$

where: $\text{T} = \text{Corporate Tax Rate}$

Analyzing Profitability

LG 5

Measures of profitability relate the returns of the firm to its sales, assets, equity, or share value. These measures allow the analyst to evaluate the firm's earnings with respect to a given level of sales, a certain level of assets, the owners' investment, or share value.

Common-size income statements express each item in the income statement as a percentage of sales. Common-size income statements are useful when comparing the performance of a firm for a particular year with that of another year.

The gross profit margin measures the percentage of each sales dollar remaining after the firm has paid for its goods.

$$\text{Gross Profit Margin} = \frac{\text{Sales} - \text{Cost of Goods Sold}}{\text{Sales}} = \frac{\text{Gross Profits}}{\text{Sales}}$$

The operating profit margin measures the percentage of profit earned on each sales dollar before interest and taxes.

$$\text{Operating Profit Margin} = \frac{\text{Operating Profits}}{\text{Sales}}$$

The net profit margin measures the percentage of each sales dollar remaining after all expenses, including taxes, have been deducted.

$$\text{Net Profit Margin} = \frac{\text{Earnings Available for Common Stockholders}}{\text{Sales}}$$

The return on total assets (ROA) measures the overall effectiveness of management in generating profits with its available assets. The ROA is also called the return on investment.

$$\text{Return on Total Assets} = \frac{\text{Earnings Available for Common Stockholders}}{\text{Total Assets}}$$

The return on equity measures the return earned on the owners' (and common stockholders') investment in the firm.

$$\text{Return on Common Equity} = \frac{\text{Earnings Available for Common Stockholders}}{\text{Common Stockholders' Equity}}$$

The earnings per share represent the number of dollars earned on behalf of each outstanding share of common stock.

$$\text{Earnings per Share} = \frac{\text{Earnings Available for Common Stockholders}}{\text{Number of Shares of Common Stock Outstanding}}$$

The price/earnings (*P*/*E*) ratio reflects the amount that investors are willing to pay for each dollar of earnings. The higher the *P*/*E* ratio, the higher the investor confidence in the firm.

$$P/E \text{ Ratio} = \frac{\text{Market Price Per Share of Common Stock}}{\text{Earnings Per Share}}$$

The market-to-book ratio reflects the level of return on equity and the degree of investor confidence.

Complete Ratio Analysis

LG 6

Complete ratio analysis includes a large number of liquidity, activity, debt, and profitability ratios. There are two popular approaches to complete ratio analysis: DuPont system of analysis and summary analysis.

1. The DuPont system of analysis merges the income statement and the balance sheet into two summary measures of profitability: return on total assets (ROA) and return on equity (ROE).

$$\text{ROA} = \frac{\text{Earnings Available for Common Stockholders}}{\text{Sales}} \times \frac{\text{Sales}}{\text{Total Assets}} = \frac{\text{Earnings Available for Common Stockholders}}{\text{Total Assets}}$$

$$\text{ROE} = \frac{\text{Earnings Available for Common Stockholders}}{\text{Total Assets}} \times \frac{\text{Total Assets}}{\text{Common Stock Equity}} = \frac{\text{Earnings Available for Common Stockholders}}{\text{Stockholders' Equity}}$$

2. A firm's performance should not be judged on a single ratio but rather on groups of ratios. To fully evaluate a corporation, four aspects need to be analyzed on a cross-sectional and time-series basis: liquidity, activity, debt, and profitability.

■ Sample Problems and Solutions

Use the balance sheet and income statements for Pizzas by Mail, Inc., that follow on this and the next page to answer the example questions.

Example 1. Basic Ratio Calculation

Calculate the ratios indicated using the financial statements that follow.

Ratio	Answer	Ratio	Answer
Net Working Capital		Debt-Equity Ratio	
Current Ratio		Times Interest Earned	
Quick Ratio		Fixed Payment Coverage	
Inventory Turnover		Gross Profit Margin	
Average Collection Period		Operating Profit Margin	
Average Payment Period		Net Profit Margin	
Fixed Asset Turnover		Return on Assets	
Total Asset Turnover		Return on Equity	
Debt Ratio		Earnings per Share	

Pizzas by Mail, Inc.
Income Statement
Year Ending December 31, 2015

Net Sales Revenue		$30,000,000
Less: Cost of Goods Sold		21,000,000
Gross Profits		9,000,000
Less: Operating Expenses:		
Selling Expense	$2,500,000	
General and Administration Expense	1,500,000	
Depreciation Expense	1,000,000	
Total Operating Expense		$ 5,000,000
Operating Profits		$ 4,000,000
Less: Interest Expense		2,000,000
Net Profits before Tax		$ 2,000,000
Less: Taxes (40%)		800,000
Net Profits After Tax		$ 1,200,000

Pizzas by Mail, Inc.
Balance Sheet
December 31, 2015

Assets	
Current Assets	
Cash	$ 1,000,000
Marketable Securities	3,000,000
Accounts Receivable	12,000,000
Inventories	7,500,000
Total Current Assets	23,500,000
Gross Fixed Assets	
Land and Buildings	$11,500,000
Machinery and Equipment	20,000,000
Furniture and Fixtures	8,000,000
Total Gross Fixed Assets	$39,500,000
Less: Accumulated Depreciation	$13,000,000
Net Fixed Assets	$26,500,000
Total Assets	$50,000,000

Pizzas by Mail, Inc.
Balance Sheet
December 31, 2015 (Continued)

Liabilities and Stockholders' Equity	
Current Liabilities:	
Accounts payable	$ 8,000,000
Notes payable	8,000,000
Accruals	500,000
Total current liabilities	$16,500,000
Long-term debt (annual payments required of $800,000)	$20,000,000
Total liabilities	$36,500,000
Stockholders' equity	
Preferred stock (100,000 shares, Div = $2.00/share)	$ 2,500,000
Common stock (1 million shares @ $5.00 par)	5,000,000
Paid in capital in excess of par on common stock	4,000,000
Retained earnings	2,000,000
Total stockholders' equity	$13,500,000
Total liabilities and stockholders' equity	$50,000,000

Solution

Ratio	Answer	Ratio	Answer
Net working capital	23.5 – 16.5 = 7,000,000	Debt-equity ratio	36.5/(13.5-2.5) = 3.32
Current ratio	1.42	Times interest earned	4/2 = 2
Quick ratio	0.97	Fixed payment coverage	4,000,000/(2,000,000 + (800,000 + 200,000 × 1/(1 – 0.4))) = 1.09
Inventory turnover	2.80	Gross profit margin	9/30 = 0.3
Average collection period	12,000,000/ (30,000,000/365) = 146	Operating profit margin	4/30 = 0.13
Average payment period	8,000,000/(21,000,000/365) = 139	Net profit margin	(1.2 – 0.20)/30 = 0.033
Fixed asset turnover	30/26.5 = 1.13	Return on assets	(1.2 – 0.2)/50 = 0.02
Total asset turnover	30/50 = 0.6	Return on equity	13.5 – 2.5 = 11 (1.2 – 0.2)/11 = 0.091
Debt ratio	36.5/50 = 0.73	Earnings per share	1,200,000 – 200,000/ 1,000,000 = 1

Example 2. Common-Size Income Statement

Prepare a common-size income statement for Pizzas by Mail, Inc., for the year ending 2015.

Pizzas by Mail, Inc.
Income Statement
Year Ending December 31, 2015

Net sales revenue	
Less: Cost of goods sold	______
Gross profits	
Less: Operating expenses:	
Selling expense	
General and administration expense	
Depreciation expense	
Total operating expense	______
Operating profits	
Less: Interest expense	______
Net profits before tax	
Less: Taxes (40%)	______
Net profits after tax	

Solution

Pizzas by Mail, Inc.
Common Size Income Statement
Year Ending December 31, 2015

Net sales revenue		100.00%
Less: Cost of goods sold		70.00%
Gross profits		30.00%
Less: Operating expenses:		
Selling expense	8.33%	
General and administration expense	5.00%	
Depreciation expense	3.33%	
Total operating expense		16.67%
Operating profits		13.33%
Less: Interest expense		6.67%
Net profits before tax		6.67%
Less: Taxes (40%)		2.67%
Net profits after tax		4.00%

Example 3. Evaluating Ratios

Pizzas by Mail has decided that it should expand nationally despite some limited complaints that pizzas are arriving cold. It has contacted Joe Flattop at the Last Chance National Bank for a $2,000,000 loan to fund the expansion. Mr. Flattop collected the following industry information to use in evaluating the loan request.

Industry Averages for Mail Order Food Businesses	
Current ratio	1.95
Debt ratio	0.46
Debt-equity ratio	1.07
Times interest earned	7.30
Fixed payment coverage ratio	1.85

Evaluate whether or not Last Chance National should extend the loan to Pizzas by Mail.

Solution

Begin by setting up a chart to compare the critical ratio with the industry ratios. If more than one year of data is available, the last several years should be included in the chart as well.

Company vs. Industry Comparison

Ratio	Industry	Pizzas by Mail	Evaluation
Current ratio	1.95	1.42	worse than industry
Debt ratio	0.46	0.73	worse than industry
Debt-equity ratio	1.07	1.48	worse than industry
Times interest earned	7.30	2.00	worse than industry
Fixed payment coverage	1.85	1.09	worse than industry

The firm has a much higher degree of indebtedness and a much lower ability to service debt than the average firm in the industry. The firm has a low current ratio. Even Last Chance National would have to decline this loan.

Example 4. Evaluating Ratios

Use the following ratios to evaluate the health of Bogus Baked Goods.

Time Series and Cross-Section Ratios
Bogus Baked Goods

	2015		2016		2017	
Ratio	**Bogus**	**Industry**	**Bogus**	**Industry**	**Bogus**	**Industry**
Current ratio	2.00	1.90	1.78	1.85	2.05	1.95
Quick ratio	1.00	0.95	0.89	0.95	0.95	0.95
Average collection period	72 days	65 days	93 days	70 days	90 days	71 days
Average payment period	80 days	70 days	117 days	75 days	106 days	75 days
Inventory turnover	3.00	4.00	2.87	3.80	2.62	4.00
Fixed asset turnover	1.33	1.50	1.11	1.55	1.42	1.60
Total asset turnover	0.80	0.90	0.69	0.93	0.78	0.95
Debt ratio	0.25	0.27	0.28	0.25	0.26	0.25
Debt-to-equity ratio	0.50	0.48	0.57	0.49	0.55	0.49
Gross profit margin	25%	24%	21%	25%	19%	27%
Operating profit margin	15%	14.8%	12%	14.9%	8.8%	15%
Net profit margin	6.7%	7.0%	5.0%	6.7%	3.5%	6.8%
Return on equity	7.15%	8.63%	4.79%	8.31%	3.69%	8.61%
Return on assets	5.3%	6.5%	3.5%	6.5%	2.8%	6.4%
Times interest earned	9.0	8.0	5.8	8.0	5.0	8.0

Solution

There are many acceptable ways to analyze financial ratios. One way that will help keep your thoughts organized and assure that you are considering every issue is to separate the ratios into categories based on what they tell us. In other words, review the liquidity ratios first, then the activity ratios, the debt ratios, and finally the profitability ratios. Review and summarize how the firm is doing in each of the four areas, then conclude how the firm is doing overall.

Liquidity: Bogus's liquidity appears to be slightly better than the industry and relatively consistent over time. Be careful not to read too much into small deviations from norms or to read a trend into normal variations that occur over time.

Activity: The activity ratios suggest there may be some problems in the firm's management. The average collection period is increasing and is much longer than the industry average. It is possible that the firm is using its credit terms as a marketing tool. This could be easily determined by asking management. The longer payment period usually indicates that the firm is having liquidity problems, but this does not appear to be the case here. It may be due to sloppy accounting systems or may be the result of negotiated terms with suppliers. Again, management should be asked to explain. The inventory turnover, fixed asset turnover, and total asset turnover are all well below the industry average. There is no evidence of improvement over the last several years. This indicates that the inventory is being mismanaged. There may be too much or obsolete inventory.

Debt Ratio: The debt ratio suggests that the firm's level of debt is in line with others in its industry and that it has not changed significantly. The times-interest-earned ratio has a declining trend. This may be due to falling profitability.

Profitability Ratios: Many analysts think that the profitability ratios are the most important because if the firm is making a fair profit, the other ratios are inconsequential. Bogus's gross profit margin is falling, and it is well below the industry average. The other profitability ratios are below the industry average and show similar falling trends. This is a *serious* problem. There are many reasons why this may happen. There may be low-cost competitors that are forcing Bogus to lower prices, or Bogus may be finding it difficult to buy its supplies as cheaply as competitors. The reason for this problem must be found and rectified.

Overall: The firm's problems seem to result from excessive levels of inventories, accounts receivable, accounts payable, and its inability to earn sufficient profits. A trend of rising costs and expenses without corresponding increases in the selling price may explain the declining returns on sales and assets. A loan officer considering extending a loan to Bogus would need to discuss the problems identified here with management.

Example 5. DuPont Analysis

Use the ratios given for Bogus Baked Goods to perform a DuPont Analysis.

Solution

ROA = Net profit margin × Total asset turnover

ROE = $3.5\% \times 0.78 \times (1/(1 - 0.26) = 3.69\%$

Because the ROE is below the industry, the analyst will want to determine whether the problem is due to the profit-on-sales component, efficiency-of-asset-use component, or a use-of-leverage component.

Example 6.

Terri Spiro, an experienced budget analyst at XYZ Corporation, has been charged with assessing the firm's financial performance during XXX5 and its financial position at year-end XXX5. To complete this assignment, she gathered the firm's XXX5 financial statements, which follow. In addition, Terri obtained the firm's ratio values for XXX4 and XXX3, along with the XXX5 industry average ratios (also applicable to XXX3 and XXX4). These are presented in the table on the following page.

XYZ Corp. Financial Statements

	Income Statement (in Thousands)		
Fiscal Year End	**01/29/XXX5**	**01/30/XXX4**	**01/31/XXX3**
Net sales	30,762,000	36,151,000	37,028,000
Cost of goods	26,258,000	29,853,000	29,732,000
Gross profit	4,504,000	6,298,000	7,296,000
Selling, general & administrative expenses	6,544,000	7,588,000	7,366,000
Operating income	–2,040,000	–1,290,000	–70,000
Non-operating income/expense	–1,067,000	–978,000	89,000
Income before interest and tax	(3,107,000)	(2,268,000)	19,000
Interest expense	155,000	344,000	287,000
Income before tax	–3,262,000	–2,612,000	–268,000
Net income before extraordinary items	–3,262,000	–2,612,000	–268,000
Extraordinary items & discontinued operations	43,000	166,000	—
Net income	–3,219,000	–2,446,000	–268,000
Outstanding shares	519,124	503,295	486,510

XYZ Corp. Financial Statements

Balance Sheet

	Assets (in Thousands)		
Fiscal Year End	**01/29/XXX5**	**01/30/XXX4**	**01/31/XXX3**
Cash	613,000	1,245,000	401,000
Accounts receivable	664,000	800,000	1,300,000
Inventories	4,825,000	5,796,000	6,051,000
Total current assets	6,102,000	7,841,000	7,752,000
Property, plant, & equipment	4,892,000	6,093,000	6,557,000
Deposits & other assets	244,000	249,000	523,000
Total assets	11,238,000	14,183,000	14,832,000

XYZ Corp. Financial Statements
Balance Sheet (Continued)

	Liabilities (in Thousands)		
Fiscal Year End	**01/29/XXX5**	**01/30/XXX4**	**01/31/XXX3**
Accounts payable	1,248,000	89,000	2,159,000
Current long term debt	—	—	68,000
Accrued expenses	872,000	563,000	1,774,000
Total current liabilities	2,120,000	652,000	4,001,000
Long term debt	8,150,000	8,555,000	2,918,000
Non-current capital leases	623,000	857,000	943,000
Total liabilities	10,893,000	10,064,000	7,862,000
Minority interest	646,000	889,000	887,000
Common stock net	519,000	503,000	487,000
Capital surplus	1,922,000	1,695,000	1,578,000
Retained earnings	–2,742,000	1,032,000	4,018,000
Shareholders equity	345,000	4,119,000	6,970,000
Total liabilities & net worth	11,238,000	14,183,000	14,832,000

XYZ Corporation
Historical Ratios

Ratio	**Actual XXX3**	**Actual XXX4**	**Actual XXX5**	**Industry Average XXX5**
Current ratio	1.938	12.026		1.1
Quick ratio	0.43	3.14		0.3
Inventory turnover (times)	4.91	5.151		7.2
Average collection period	72.462	8.077		
Total asset turnover (times)	2.496	2.549		2.7
Debt ratio	0.53	0.71		0.44
Times interest earned ratio	0.066	–6.593		
Gross profit margin	19.70%	17.42%		24.9
Net profit margin	–0.72%	–6.77%		3.2
Return on total assets (ROA)	–0.018	–0.172		8.1
Return on common equity (ROE)	–0.038	–0.594		19.6
Price/earnings (*P*/*E*) ratio				22.9
Market/book (M/B) ratio				10.21

a. Calculate the firm's XXX5 financial ratios, and then fill in the preceding table.

b. Analyze the firm's current financial position from both a cross-sectional and a time-series viewpoint. Break your analysis into evaluations of the firm's liquidity, activity, debt, profitability, and market.

c. Summarize the firm's overall financial position on the basis of your findings in part b.

Solution

a. Ratio Calculations

Financial Ratio	XXX3
Current ratio	$6,102,000 ÷ $2,120,000 = 2.878
Quick ratio	($6,102,000 – $4,825,000) ÷ $1,277,000 = 0.60
Inventory turnover (times)	$26,258,000 ÷ $4,825,000 = 5.442
Average collection period (days)	$664,000 ÷ ($30,762,000 ÷ 365) = 7.879
Total asset turnover (times)	$30,762,000 ÷ $11,238,000 = 2.737
Debt ratio	$10,893,000 ÷ $11,238,000 = 0.969
Times interest earned	–$3,107,000 ÷ $155,000 = –20.045
Gross profit margin	(30,762,000 – 26,258,000) ÷ 30,762,000 = 14.64%
Net profit margin	–$3,219,000 ÷ $30,762,000 = –10.46%
Return on total assets	–$3,219,000 ÷ $11,238,000 = – 0.286
Return on equity	–$3,219,000 ÷ $345,000 = – 9.330

XYZ Corporation
Historical Ratios

Ratio	Actual XXX3	Actual XXX4	Actual XXX5	Industry Average XXX5
Current ratio	1.938	12.026	2.878	1.1
Quick ratio	0.43	3.14	0.60	0.3
Inventory turnover (times)	4.91	5.151	5.442	7.2
Average collection period	72.462	8.077	7.879	
Total asset turnover (times)	2.496	2.549	2.737	2.7
Debt ratio	0.53	0.71	0.969	0.44
Times interest earned ratio	0.066	–6.593	–20.045	
Gross profit margin	19.70%	17.42%	14.64%	24.9
Net profit margin	–0.72%	–6.77%	–10.46%	3.2
Return on total assets (ROA)	–0.018	–0.172	–0.286	8.1
Return on common equity (ROE)	–0.038	–0.594	–9.330	19.6
Price/earnings (*P*/*E*) ratio				22.9
Market/book (M/B) ratio				10.21

b. *Liquidity:* The firm has sufficient current assets to cover current liabilities. The firm's liquidity is higher than the industry average. The trend is downward from years XXX4 to XXX5 and getting closer to the industry average.

Activity: The inventory turnover is stable but lower than the industry average, which could indicate the firm is holding too much inventory. The average collection period is decreasing due to a decrease in accounts receivable. Total asset turnover is stable and matches the industry average. This indicates that the sales volume is sufficient for the amount of committed assets.

Debt: The debt ratio has increased and is substantially higher than the industry average.

This places the company at high risk. Typically, industries with heavy capital investment and higher operating risk try to minimize the financial risk. XYZ Corporation has positioned itself with both heavy operating and financial risk. The times-interest-earned ratio is decreasing and also indicates a potential debt service problem.

Profitability: The gross profit margin is decreasing slightly and is well below the industry average. The next profit margin is also decreasing and far below the industry average. This is an indicator that the firm does not have sufficient sales dollars remaining after expenses have been deducted. The high financial leverage has caused the low profitability.

Market: Return on equity and return on assets are both decreasing and are well below the industry average. This indicates a problem of management in generating profits with its available assets, as well as a problem with the return earned on the owner's investment in the firm.

c. XYZ Corporation has a problem with sales not being at an appropriate level for its capital investment. As a consequence, the firm has acquired a substantial amount of debt which, due to the high interest payments associated with the large debt burden, is depressing profitability. These problems may be picked up by investors and reflect in market ratios.

■ Study Tips

This chapter presents and explains the more common financial ratios used to analyze firms. A few points need to be emphasized.

1. Seldom will ratios answer questions; most of the time they raise questions. This is valuable because by raising questions they give the analyst a direction to continue study. For example, a loan officer will be able to ask the *right* questions of a loan applicant.

2. Do not get too excited over small annual deviations from the industry or from historical values. This is natural to businesses. Look for trends that indicate a problem that is more than just an annual anomaly.

3. Do not put too much faith in one ratio by itself. There are several ratios in each major area of analysis. Use them all and view all of them together to get the big picture.

■ Student Notes

■ Sample Exam—Chapter 3

True/False

T F 1. The more financial leverage that a firm uses, the greater will be its risk and expected return.

T F 2. Scantron Corporation's inventory turnover ratio is twice as fast as the industry average. It is safe to assume that Scantron is a profitable corporation.

T F 3. The creditors of a firm must be satisfied before any earnings can be distributed to the common shareholders.

T F 4. Common-size income statements restate each item in the statement as a percentage of net income.

T F 5. The operating profit margin must take into account interest and taxes.

T F 6. A *P*/*E* ratio of 20 indicates that investors are willing to pay $20 for each $1 of earnings.

T F 7. Earnings per share is calculated by dividing retained earnings by the number of shares of common stock outstanding.

T F 8. Return on total assets (ROA) is sometimes called the return on investment.

T F 9. Liquidity ratios are used to measure the speed with which various accounts are converted into sales.

T F 10. Generally, inventory is considered the most liquid asset that a firm possesses.

T F 11. The current ratio measures the firm's ability to meet short-term obligations.

T F 12. When referring to ratio comparisons, time-series analysis compares a firm to that of an industry leader.

T F 13. When ratios of different years are being compared, inflation should be taken into consideration.

T F 14. Standard Corporation reported a gross profit margin of 28 percent in 2015. Parker Inc. reported a gross profit margin of 15 percent in 2015. It is safe to assume that Standard Corporation generated higher operating profits than Parker in 2015.

T F 15. The DuPont system of analysis merges a firm's income statement and balance sheet into a summary measure of profitability.

T F 16. Personal financial statements are necessary for the establishment and monitoring of your progress towards personal financial goals.

T F 17. Generally Accepted Accounting Principles (GAAP) are the procedure guidelines used to prepare and maintain financial records and reports and are authorized by the Financial Accounting Standards Board (FASB).

T F 18. Benchmarking is a type of cross-sectional analysis in which the firm's ratio values are compared with those of a key competitor or with a group of competitors that it wishes to emulate.

Multiple Choice

1. Carter Corporation has current assets of $120 million and inventory equal to $30 million. If Carter's current liabilities are $100 million, what will the current ratio be?

   a. 0.90
   b. 1.20
   c. 4
   d. 1.71

2. An increased debt position will be accompanied by __________ risk.

   a. undetermined
   b. unchanged
   c. less
   d. greater

3. __________ refers to the overall solvency of the firm—the ease with which it can pay its bills.

   a. Liquidity
   b. Turnover
   c. Leverage
   d. Coverage

4. Stanton Inc. reported annual sales of $400,000 in 2012. At year end, the balance in accounts receivables was reported to be $10,000. What was the average collection period for Stanton based on a 360 day year?

   a. 2.5 days
   b. 40 days
   c. 4 days
   d. 9 days

5. __________ are used to measure the speed in which various accounts are converted into sales or cash.

   a. Liquidity ratios
   b. Activity ratios
   c. Debt ratios
   d. Profitability ratios

6. A company's fixed assets are termed its ___________ assets.
   a. short-term
   b. earning
   c. financed
   d. equity

7. The _________ ratio measures the financial leverage of the firm.
   a. current
   b. times interest earned
   c. debt-equity
   d. acid test

For question 8, refer to the following information.

ABC Corporation	
Sales Revenue	XX%
Less: Cost of Goods Sold	XX%
Gross Profit Margin	25%

8. What will ABC's cost of goods sold be if expressed as a percentage of sales?
   a. 125 percent
   b. 25 percent
   c. 100 percent
   d. 75 percent

9. _______ is/are generally the least liquid current asset that a corporation possesses.
   a. Marketable securities
   b. Cash
   c. Inventory
   d. Accounts receivable

10. ________ involves comparing a firm to the industry leader or to an industry average.
   a. Coverage analysis
   b. Cross-sectional analysis
   c. DuPont analysis
   d. Time-series analysis

11. Inventory values and asset values can differ year to year due to
   a. inflation.
   b. increased cost of capital.
   c. recessions.
   d. depressions.

12. Net working capital is calculated as current assets minus
    a. inventory.
    b. cost of goods sold.
    c. fixed assets.
    d. current liabilities.

13. A firm with sales of $500,000, net profits after taxes of $20,000, total liabilities of $200,000, and stockholders' equity of $100,000 will have a return on equity of
    a. 5 percent.
    b. 20 percent .
    c. 10 percent .
    d. 40 percent .

14. The ______ ratio indicates the amount of money that investors are willing to pay for $1 of earnings.
    a. EPS
    b. times interest earned
    c. *P/E*
    d. earnings

15. The DuPont system of analysis allows firms to break down their return on equity down into all of the following components EXCEPT
    a. inventory usage.
    b. profit in sales.
    c. efficiency of asset usage.
    d. use of leverage.

16. Starbuck Corporation reported EPS of $2.30 for 2012. In 2012 Starbuck had earnings available to common stockholders of $1,380,000. How many outstanding shares of common stock did Starbuck have in 2012?
    a. 3,174,000
    b. 600,000
    c. 400,000
    d. 3,600,000

17. The __________ ratio measures the firm's ability to meet payment obligations such as loan interest and principal, and preferred stock dividends.
    a. times-interest-earned
    b. debt-equity
    c. current ratio
    d. fixed payment coverage

18. In the ________ income statement, each item is expressed as a percentage of sales.
    a. pro forma
    c. common-size
    b. second stage
    d. sales-based

For questions 19 and 20, refer to the following information.

Reeves Enterprises
Year Ended December 31, 2015

Sales	$ 400,000	Stockholder's Equity	$200,000
Total Assets	1,000,000	Cost of Goods Sold	100,000
Total Liabilities	900,000	Net Profit After Taxes	70,000
EPS	4.3	Long-term Debt	400,000

19. What will Reeves' ROA be for 2015 under the DuPont system?
   a. 7 percent
   b. 10 percent
   c. 4 percent
   d. 14 percent

20. What will Reeves' ROE be for 2015 under the DuPont system?
   a. 14 percent
   b. 35 percent
   c. 18 percent
   d. 10 percent

21. What is the net worth of Carl and Carol Luedtke, a retired couple who live in their RV on rented property in Florida, have the following assets and liabilities.

Item	Amount
Checking account	$ 1,600
Money market fund	$ 2,200
Stocks	$ 22,000
Bonds	$ 19,000
2008 Nissan Altima	$ 23,000
2001 Winnebago RV	$134,000
Furnishings	$ 2,100
Jewelry	$250
Bank card balance	$880
Unpaid utility bill	$170
Auto loan	$ 26,000

   a. $20,100
   b. $43,100
   c. $177,100
   d. $231,200

22. Which of the following is not considered when calculating the personal liquidity ratio?
    a. annual mortgage payments
    b. bank card balances
    c. auto loan balances
    d. savings account balances

23. The statement that provides a summary of the firm's operating, investment, and financing cash flows and reconciles them with changes in cash and marketable securities during the period is:
    a. The income statement
    b. The balance sheet
    c. The statement of shareholder equity
    d. The statement of cash flows.

24. Given long-term debt of $2.1 million, total liabilities of $5.3 million, preferred equity of $1.0 million and total equity of $4.5 million. What is the debt-to –equity ratio of the firm?
    a. 047
    b. 1.18
    c. 1.51
    d. 0.60

Essay

1. Explain why financial leverage is associated with risk.

2. List some of the precautions that should be taken when performing ratio analysis.

■ Chapter 3 Answer Sheet

True/False

1. T
2. F
3. T
4. F
5. F
6. T
7. F
8. T
9. F
10. F
11. T
12. F
13. T
14. F
15. T
16. T
17. T
18. T

Multiple Choice

1. B — $\frac{\$120\text{M}}{\$100\text{M}} = 1.2$
2. D
3. A — $\frac{\$10{,}000}{(\$400{,}000/360)} = 9 \text{ days}$
4. D
5. B
6. B
7. C — Sales = 100%; COGS = 100% – 25% = 75%
8. D
9. C
10. B
11. A
12. D — $\text{ROE} = \frac{\$20{,}000}{\$100{,}000} = 20\%$
13. B
14. C
15. A — $\$2.30 = \frac{\$1{,}380{,}000}{\text{Shares of Common}}$; Shares of Common = 600,000
16. B
17. D
18. C — $\text{ROA} = \left(\frac{\$70{,}000}{\$400{,}000}\right) \times \left(\frac{\$400{,}000}{\$1{,}000{,}000}\right)$; ROA = 7%
19. A
20. B — $\text{ROE} = \left(\frac{\$70{,}000}{\$1{,}000{,}000}\right) \times \left(\frac{\$1{,}000{,}000}{\$200{,}000}\right)$; ROE = 35%
21. C
22. C
23. D
24. C — Debt-to- Equity = 5.3/(4.5 – 1) = 1.51

Essay

1. Financial leverage is a term that is used to describe the ratio of debt financing to the stockholders' equity. Debt financing is considered to carry a greater risk than equity financing because creditors must be satisfied before any earnings can be distributed to the stockholders. The larger the amount of debt to equity that a firm carries, the greater the chance that the common shareholders will not receive any of the firm's earnings. Because of this risk associated with debt financing, a firm that is heavily leveraged is considered to be risky.

2. a. Do not use a single ratio to judge the overall performance of the firm.
   b. The financial statements being compared should be dated at the same point of time during the year.
   c. Audited financial data should be used to ensure relevant financial information.
   d. The financial data being compared in the ratio analysis should be developed in the same manner.
   e. When performing time-series analysis, inflation should always be taken into account.

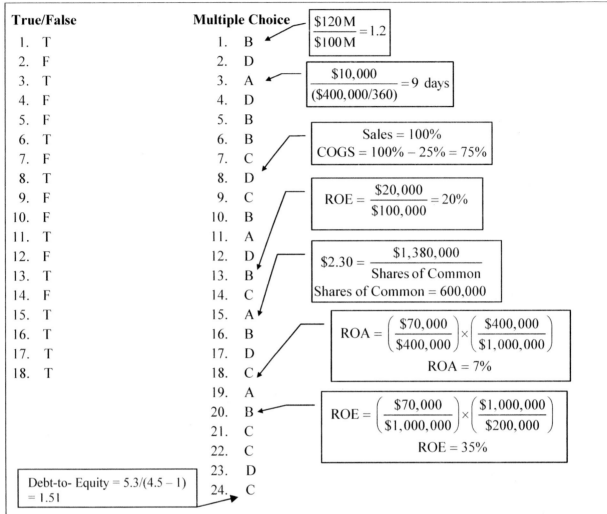

Chapter 4
Cash Flow and Financial Planning

■ Chapter Summary

Financial planning is an important part of every financial manager's job. By accurately projecting future cash surpluses and shortfalls, timely arrangements can be made for investing or borrowing needs. For example, bankers will look much more favorably on a loan request to fund a temporary cash need anticipated two months in the future than it will on a loan request for funds needed right now. Two methods are provided to compute the external funds needed by a firm.

Accurate financial planning is difficult because of the need to project uncertain future events. For example, a developer may not need cash if his properties sell but may be in deep trouble if they do not. Financial planning under uncertainty is addressed in this chapter.

LG 1 *Understand tax depreciation procedures and the effect of depreciation on the firm's cash flows.* Depreciation represents the systematic charging of a portion of the costs of fixed assets against annual revenues over time. A firm will often use different depreciation methods for financial reporting than for tax calculations. For tax calculations, corporations use MACRS. When estimating cash flows, depreciation and any other noncash charges are added back to operating profits. Under the MACRS procedure, the depreciable value of an asset is its full cost, including outlays for installation. No adjustment is required for expected salvage value. The MACRS procedures classify assets other than real estate into property classes based on recovery periods of 3, 5, 7, 10, 15, and 20 years. Each recovery period has a schedule of depreciation percentages for each year of the recovery period.

LG 2 *Discuss the firm's statement of cash flows, operating cash flow, and free cash flow.* The statement of cash flows summarizes the firm's cash flow over a given period of time. Managers must identify all corporate sources and uses of funds and then classify these items as operating, investment, or financing cash flows. The statement of cash flows allows financial managers to analyze and explain the firm's sources and uses of cash over time. Because the focus of finance is on cash, free cash flow, the amount of cash flow available to creditors and owners, is of great importance.

LG 3 *Understand the financial planning process, including long-term (strategic) financial plans and short-term (operating) financial plans.* There are two key elements to the financial planning process, cash planning and profit planning. Cash planning involves preparation of the firm's cash budget. Financial planning begins with the preparation of a long-term strategic plan that guides the preparation of short-term operating plans. Profit planning relies on the pro forma income statement and balance sheet.

LG 4 *Discuss the cash-planning process and the preparation, evaluation, and use of the cash budget.* The cash budget is a tool for forecasting the firm's cash flow in the future. The cash budget is even important to firms that have surplus cash because it will let them take advantage of higher returns on longer term investments. Cash budgets usually begin with a sales forecast. Projected disbursements are subtracted from the receipts to arrive at the ending cash balance. The cash balance is rolled forward from period to period. Once the cash budget has been prepared, it must be interpreted. Most cash budgets represent a best effort to project the future. But because it involves a projection, there will usually be an element of uncertainty. Some firms choose to compute best-case, most-likely, and worst-case scenarios. Others use computer simulation to deal with this uncertainty.

LG 5 *Explain the simplified procedures used to prepare and evaluate the pro forma income statement and the pro forma balance sheet.* The judgment approach to pro forma balance sheet and income statement preparation involves estimating some account values and calculating others, thereby requiring a "plug" figure to bring the statement into balance. If a plug figure must be added to the right side of the balance sheet, additional external funds are needed.

LG 6 *Evaluate the simplified approaches to pro forma financial statement preparation and the common uses of pro forma statements.* Pro forma statements are based on the assumption that the firm's past financial condition is an accurate indicator of its future and that certain variables, such as cash, accounts receivable, and inventories, can be forced to take on certain desired values. These assumptions may not always reflect the truth, and the financial manager must be careful to monitor the accuracy of the projections.

■ Chapter Notes

Depreciation

LG 1 Business firms are allowed to systematically charge a portion of the costs of fixed assets against revenues. This is called depreciation. Depreciation lowers the tax base and is beneficial to corporations from a tax standpoint. Depreciation is computed using straight line or MACRS. To compute straight-line depreciation, add together all costs associated with putting the asset into service, subtract the salvage value, and then divide by the expected life of the asset.

Depreciation for tax purposes is determined by the modified accelerated cost recovery system (MACRS).

Remember from Chapter 1 that the financial manager is only concerned with cash flows rather than the operating profit that is reported in the income statement. To adjust the income statement to show cash flows from operation, all noncash charges must be added back to the firm's net profits. These noncash charges include depreciation, amortization, and depletion allowances.

Under the MACRS procedure, an asset is depreciated at its full cost, including outlays for installation. No adjustment for salvage value is needed.

Under MACRS, an asset's depreciable life is broken down into six property classes. The recovery periods are 3, 5, 7, 10, 15, and 20 years. Assets such as research equipment and specialty tools would fall into the 3-year recovery class. More expensive items such as buildings would fall into the 20-year recovery class.

Compute depreciation by multiplying the original asset value by the percentage listed for each year under the asset's class life. For example, if the asset has a five-year class life and originally cost \$10,000, the first year's depreciation is computed by multiplying \$10,000 × 0.20 = \$2,000.

Cash Flows

LG 2

A firm's cash flow can be broken down into three divisions:

1. *Operating flows* are cash flows directly related to the production and sale of the firm's products.
2. *Investment flows* are cash flows associated with the purchase and selling of both fixed assets and business interest.
3. *Financing flows* are cash flows resulting from the debt and equity financing transactions.

The statement of cash flows summarizes the sources and uses of cash for the firm. For example, an increase in accounts payable by $100 is a source of cash. An increase in a firm's inventory by $200 is a use of cash because an additional $200 would be tied up in the firm's inventory.

The **statement of cash flows** can be developed in a five-step process:

1. Calculate the changes in assets, liabilities, and stockholders' equity.
2. Determine if the change is a source or a use of funds.
3. Separately sum all the sources and uses of funds. If done correctly, these figures should be equal.
4. Net profit after taxes, depreciation, and any other noncash charges, and dividends need to be determined. Depreciation and the other noncash charges can then be added to the net operating profit after taxes in the cash flows from operations, and the dividends can be taken out of the cash flows from financing activities.
5. The relevant data from steps 1, 2, 3, and 4 need to be broken down into cash flows from operating activities, investing activities, or financing activities. Once the five steps are completed, the cash flow statement can be prepared.

Uses and Sources of Funds Statement: To prepare the uses and sources of funds statement, evaluate each balance sheet line item to determine whether it is a source of cash or a use of cash. Sources and uses of cash are determined as follows:

Sources: In flow

1. Decrease in an asset account
2. Increase in a liability account

Uses: Outflow

1. Increase in an asset account
2. Decrease in a liability account

Free cash flows represent the amount of cash flow available to investors—the providers of debt and equity—after the firm has met all operating needs and paid for investments in net fixed assets and net current assets.

Financial Planning

LG 3

Financial planning guides the actions of a firm toward the achievement of immediate and long-term goals. The financial planning process begins with long-term, or strategic, financial plans that in turn guide the formulation of short-term, or operating, plans and budgets.

Long-term (strategic) financial plans: Long-term financial plans are planned long-term financial actions and the anticipated financial impact of those actions. These plans tend to cover periods of 2 to 10 years and typically deal with such things as proposed fixed-asset outlays, research and development activities, and major sources of financing.

Short-term (operating) financial plans: Short-term financial plans are short-term financial actions and the anticipated financial impact of those actions. These plans tend to cover a 1- to 2-year period and are centered on key inputs, which include sales forecasts and various forms of operating and financial data, and key outputs, which include a number of operating budgets, the cash budget, and pro forma financial statements.

Cash Planning: Cash Budgets

LG 4

The cash budget allows the firm to forecast its short-term cash requirements, typically during a year, divided into monthly intervals. In planning the cash budget, the firm pays particular attention to surplus cash and cash shortages.

Sales Forecasts

The main component in the financial planning process, and therefore any cash budget is the firm's sales forecast. Based on the sales forecast, the financial manager estimates the monthly cash flows that will result from projected sales receipts and from production-related, inventory-related, and sales-related outlays. The financial manager also determines the level of fixed assets required and the amount of financing needed to support the forecasted level of sales. The sales forecast may be based on an analysis of external or internal data or a combination of both.

External forecast: An external forecast is based on the relationship that can be observed between the firm's sales and certain key external economic indicators, such as the gross domestic product (GDP), new housing starts, and disposable personal income.

Internal forecast: Internal forecasts are based on a consensus of forecasts made by the firm's salespeople in the field.

Combined forecast: Firms most often use a combination of external and internal forecasts. The internal data provides insight into sales expenditures, and the external data provides a means of adjusting these expenditures to take into account general economic factors.

Preparing the Cash Budget

The general format for the cash budget is presented in the following table.

	Jan.	Feb.	…	Nov.	Dec.
Cash receipts			…		
Less: Cash disbursements	_____	_____	…	_____	_____
Net cash flow			…		
Add: Beginning cash	_____	_____	…	_____	_____
Ending cash	↗	↗	… ↗	↗	
Less: Minimum cash balance	_____	_____	…	_____	_____
Required total financing			…		
Excess cash balance			…		

Cash receipts: These include all items from which cash inflows result in any given financial period. Examples are cash sales, collection of accounts receivable, and other cash receipts.

Cash disbursements: These include all outlays of cash in the period covered. Examples are cash purchases, payment of accounts payable, rent payments, and tax payments.

Net cash flow, ending cash, financing, and excess cash: A firm's net cash flow is found by subtracting the cash disbursements from cash receipts in each period. By adding beginning cash to the firm's net cash, the ending cash for each period can be found. Finally, subtracting the desired minimum cash balance from ending cash yields the required total financing or the excess cash balance.

Evaluating the Cash Budget

The cash budget can determine whether an excess of cash or a shortage of cash will result in each period. Excess cash can be invested in marketable securities, and shortages of cash can be financed with short-term borrowing (notes payable).

In evaluating the cash budget, a level of uncertainty does exist. One way to cope with this uncertainty is for the financial manager to prepare a best-case forecast, a most-likely forecast, and a worst-case forecast. An evaluation of these cash flows allows the financial manager to determine the amount of financing necessary to cover the most adverse situations. A second way that a financial manager can cope with uncertainty when preparing a cash budget is to use computer simulation. By simulating the occurrences of sales, the firm can develop a probability distribution of its ending cash flows for each month.

Profit Planning: Pro Forma Statement Fundamentals

LG 5

Profit planning typically involves the preparation of a pro forma (projected) income statement and balance sheet. The two key inputs that are required for preparing pro forma statements are (1) the financial statements from the preceding year and (2) the sales forecast for the coming year.

Pro Forma Income Statement

A simple method for developing a pro forma income statement is to use the percent-of-sales method. This method for preparing the pro forma income statement assumes that the relationship between expenses and sales in the coming year will be identical to those in the immediately preceding year.

Example

The following percent-of-sales figures were taken from Encore Inc.'s 2011 income statement:

$$\frac{\text{COGS}}{\text{Sales}} = 75\%$$

$$\frac{\text{Operating Expenses}}{\text{Sales}} = 15\%$$

$$\frac{\text{Interest Expense}}{\text{Sales}} = 2\%$$

Assuming the relationship between expenses and sales is the same for 2012 as for 2011, Encore's pro forma income statement for 2012 based on sales is as follows:

Sales revenue	$160,000
Less: Cost of goods sold (0.75)	120,000
Gross profits	$ 40,000
Less: Operating expenses (0.15)	24,000
Operating profits	$ 16,000
Less: Interest expense (0.02)	3,200
Net profit before taxes	$ 12,800
Less: Taxes (0.40 × 12,800)	5,120
Net profits after taxes	$ 7,680

A strict application of the percent-of-sales method assumes that all costs are variable. Nearly all firms have fixed costs, and ignoring this in the pro forma income statement usually results in the misstatement of the firm's forecast profit. It is advisable when using the percent-of-sales method to break down costs into fixed and variable components. Fixed costs do not vary with differing levels of sales and should be estimated by the financial manager based on upcoming capital expenditures and historic fixed costs.

Pro Forma Balance Sheet

The most popular method for developing a pro forma balance sheet is the judgmental approach. This approach estimates the values of certain balance sheet accounts while calculating others and uses the firm's external financing as a balancing figure. A positive value for the balancing figure, which is stated on the balance sheet as "external funds required," indicates that the firm must raise external funds through debt and/or equity in order to support the firm's forecasted level of sales. If a negative value exists for external funds required, the firm's forecasted financing is in excess of its needs.

Evaluation of Pro Forma Statements

LG 6

The percent-of-sales method and judgmental approach to pro forma preparation are popular because of their simplicity, but they do contain weaknesses.

The weakness of the simple pro forma approaches lies within two of their basic assumptions:

1. The firm's past financial condition is an accurate indicator of its future financial performance.
2. Certain variables, such as cash, accounts receivable, and inventories, can be forced to take on certain "desired" levels.

■ Sample Problems and Solutions

Example 1. Depreciation

Earlier this year, Murphy Corporation purchased a new machine to package their salsa sauce. The machine cost $22,000, and there were installation costs of $2,000. The machine has a five-year recovery period and is expected to have a salvage value of $3,000. Develop a depreciation schedule for this asset using the MACRS depreciation percentages given below.

	Percentage by recovery year
Recovery year	5 years
1	20%
2	32%
3	19%
4	12%
5	12%
6	5%
Totals	100%

Prepare your solution in the table below:

Depreciation Schedule Packaging Machine			
Year	**Cost**	**MACRS Depreciation Percentages**	**Depreciation**
1			
2			
3			
4			
5			
6			
Total			

Solution

Depreciation Schedule Packaging Machine			
Year	**Cost (1)**	**MACRS Depreciation Percentages (2)**	**Depreciation [(1) × (2)] = (3)**
1	24,000	20%	4,800
2	24,000	32%	7,680
3	24,000	19%	4,560
4	24,000	12%	2,880
5	24,000	12%	2,880
6	24,000	5%	1,200
Total			24,000

Notes: The total depreciable cost is the cost of the machine, $22,000, plus the installation costs of $2000. The salvage value is not used in the MACRS calculation of depreciation. The sum of the depreciation for each of the six years is equal to the total depreciable cost.

Example 2. Cash Budgets

Prepare a cash budget for the first quarter for With-It Wonderwear, Inc., given the following data.

Management has projected total and credit sales for the first three months of 2014 as follows:

Month	Year	Total Sales	Credit Sales
December	2013	$825,000	$770,000
January	2014	730,000	690,000
February	2014	840,000	780,000
March	2014	920,000	855,000

Based on historical data, 25 percent of the credit sales are collected in the month the sales were made, and 75 percent are collected the next month. Purchases are estimated to be 60 percent of the next month's total sales but are paid for in the month following the order. This means that the expenditure for inventory in each month is 60 percent of that month's sales. The beginning and target cash balance is $100,000. Other disbursements are as follows:

Expense Item	January	February	March
Wages	$250,000	$290,000	$290,000
Rent	27,000	27,000	27,000
Other Expense	10,000	12,000	14,000
Taxes	105,000		
Dividends			40,000
Capital Expenditures		75,000	

Prepare your solution on the table below:

CASH BUDGET

	January	February	March
SALES	730,000	840,000	920,000
Credit Sales	690,000	780,000	855,000
RECEIPTS			
Cash sales			
Collections			
Total Cash			
DISBURSEMENTS			
Payments of A/P			
Wages			
Rent			
Other Expense			
Taxes			
Dividends			
Cap Expense			
TOTAL EXPENSE			
Begin Cash			
CASH IN-OUT			
Ending Cash			
TARGET BALANCE			
LOAN REQUIRED			

Solution

CASH BUDGET

	January	February	March
SALES	730,000	840,000	920,000
Credit Sales	690,000	780,000	855,000
RECEIPTS			
Cash sales	40,000	60,000	65,000
Collections	750,000	712,500	798,750
Total Cash	790,000	772,500	863,750
DISBURSEMENTS			
Payments of A/P	438,000	504,000	552,000
Wages	250,000	290,000	290,000
Rent	27,000	27,000	27,000
Other Expense	10,000	12,000	14,000
Taxes	105,000		
Dividends			40,000
Cap Expense		75,000	
TOTAL EXPENSE	830,000	908,000	923,000
Begin Cash	100,000	100,000	100,000
CASH IN-OUT	(40,000)	(135,500)	(59,250)
Ending Cash	60,000	(35,500)	40,750
TARGET BALANCE	100,000	100,000	100,000
LOAN REQUIRED	(40,000)	(135,500)	(59,250)

730,000 – 690,000

(0.25 × 690,000) + (0.75 × 770,000)

0.60 × 730,000

Total cash receipts – total expense

(100,000 – 60,000) The amount needed to bring the balance up to 100,000.

Note: Cash budgets tend to be varied due to the different types of projections available and the needs of the firm. The cash budget presented here is similar to, but not exactly like, the one presented in the text. You may have developed a still different model. The important concept is to estimate total cash receipts, and outflow and to determine how much additional cash will be required to reach the target cash balance.

Example 3. Scenario Analysis

Joe Blow has invented a toe exerciser that he expects to sell by mail-order. He projects sales of $200,000 during each of the months of January, February, and March. His monthly purchases during this time will be $140,000, wages and salaries will be $20,000 per month, and a monthly rent of $5,000 must be paid. Taxes of $30,000 are due in January. A $25,000 payment for the purchase of capital equipment is due in February. All sales are for cash, and beginning cash balances are assumed to be zero. Joe wants to maintain a $5,000 minimum cash balance. While Joe is certain of all other figures, he is a little nervous about the accuracy of his sales forecast. He feels a pessimistic sales figure of $180,000 per month and an optimistic figure of $220,000 is accurate. What are the minimum and maximum cash balances the company can expect for each of the next three monthly periods?

Solution

Summary of Disbursements			
	January	**February**	**March**
Purchases	$140,000	$140,000	$140,000
Wages	20,000	20,000	20,000
Rent	5,000	5,000	5,000
Taxes	30,000		
Capital Equipment		25,000	
Total Cash Disbursements	$195,000	$190,000	$165,000

	January			**February**			**March**		
Cash Flow	**Pessimistic**	**Most Likely**	**Optimistic**	**Pessimistic**	**Most Likely**	**Optimistic**	**Pessimistic**	**Most Likely**	**Optimistic**
Total Cash Receipts	$180	$200	$220	$180	$200	$220	$180	$200	$220
Less: Total Cash Disbursements	195	195	195	190	190	190	165	165	165
Net Cash Flow	($ 15)	$ 5	$ 25	($ 10)	$ 10	$ 30	$ 15	$ 35	$ 55
Add: Beginning Cash	0	0	0	($ 15)	$ 5	$ 25	($ 25)	$ 15	$ 55
Ending Cash	$(15)	$ 5	$ 25	($ 25)	$ 15	$ 55	($ 10)	$ 50	$110
Less: Minimum Cash Balance	5	5	5	5	5	5	5	5	5
Required Total Financing	$ 20			$ 30			$ 15		
Excess Cash Balance			$ 20		$ 10	$ 50		$ 45	$105

During January, Joe Blow will need a maximum of $20,000 of financing, while at best, he will have $20,000 available for short-term investment. During February, the firm's maximum financing requirement will be $30,000; however, it could experience a surplus between $10,000 and $50,000. The March projections reflect a borrowing requirement of $15,000, with a surplus between $45,000 and $105,000. By considering the extreme values reflected in the pessimistic and optimistic outcomes, Joe should be able to plan his cash requirements more accurately.

Example 4. Pro Forma Financial Statements

The Kiddy Car Driving School wishes to use the percent-of-sales shortcut method to prepare their 2015 pro forma financial statements. Use the operating statements below and the additional information given to prepare a pro forma income statement and balance sheet.

1. The firm has estimated that its sales for 2015 will be $1,350,000.
2. The firm expects to pay $52,000 in cash dividends in 2015.
3. Taxes payable will equal 25 percent of the tax liability on the pro forma income statement.
4. All assets and all current liabilities change as a percentage-of-sales.

The balance sheet and income statement needed to complete this problem appear on the next page.

Prepare your solution here.

Kiddy Car Driving School **Pro Forma Income Statement** **January 2015–December 2015**
Net Revenues
Less Cost of Goods Sold
Gross Profit
Less: Operating Expenses
Profits before Taxes
Less: Taxes (40%)
Net Profits after Taxes
Less: Cash Dividend
Increase in Retained Earnings

Kiddy Car Driving School
Pro Forma Balance Sheet
December 31, 2015

Assets	**Liabilities and Owners' Equity**
Cash	Accounts Receivable
Marketable Securities	Taxes Payable
Accounts Receivable	Other Current Liabilities
Inventory	Total Current Liabilities
Total Current Assets	Long-Term Debt
Net Fixed Assets	Common Stock
	Retained Earnings
Total Assets	Total Liabilities and Equity

Kiddy Car Driving School
Income Statement
January 2014–December 2014

Net Revenues	$1,200
Less Cost of Goods Sold	900
Gross Profit	300
Less: Operating Expenses	150
Profits before Taxes	150
Less: Taxes (40%)	60
Net Profits after Taxes	90
Less: Cash Dividend	30
Increase in Retained Earnings	$ 60

Kiddy Car Driving School
Balance Sheet
December 31, 2014

Assets		**Liabilities and Owners' Equity**	
Cash	$ 48	Accounts Receivable	$156
Marketable Securities	36	Taxes Payable	15
Accounts Receivable	216	Other Current Liabilities	0
Inventory	144	Total Current Liabilities	171
Total Current Assets	444	Long-Term Debt	300
Net Fixed Assets	540	Common Stock	225
		Retained Earnings	288
Total Assets	$984	Total Liabilities and Equity	$984

Solution

Kiddy Car Driving School
Pro Forma Income Statement
January 2015–December 2015

Net Revenues	$1,350.00
Less Cost of Goods Sold	1,012.50
Gross Profit	337.50
Less: Operating Expenses	168.75
Profits before Taxes	168.75
Less: Taxes (40%)	67.50
Net Profits after Taxes	101.25
Less: Cash Dividend	52.00
Increase in Retained Earnings	$ 49.25

900/1,200 = 0.75
0.75 × 1,350 = 1,012.50

150/1,200 = 0.125
0.125 × 1,350 = 168.75

Given in problem

Add this figure to the 2004 retained earnings to find the 2005 retained earnings.

Kiddy Car Driving School
Pro forma Balance Sheet
December 31, 2015

Assets		**Liabilities and Owners' Equity**	
Cash	$ 54.00	Accounts Receivable	$ 175.50
Marketable Securities	40.50	Taxes Payable	16.87
Accounts Receivable	243.00	Other Current Liabilities	0.00
Inventory	162.00	Total Current Liabilities	192.37
Total Current Assets	499.50	Long-Term Debt	300.00
Net Fixed Assets	607.50	Common Stock	225.00
		Retained Earnings	337.25
		External Funds Required	52.37
Total Assets	$1,107.00	Total Liabilities and Equity	$1,107.00

48/1,200 = 0.04
0.04 × 1,350 = 54
Each balance is found in the same way.

Only current assets and liabilities are assumed to fluctuate with sales.

Plug external financing required to balance total liabilities and equity to total assets.

Based on the pro forma statements, the firm will have an external funding requirement of $52,375 to support its planned 2015 sales level.

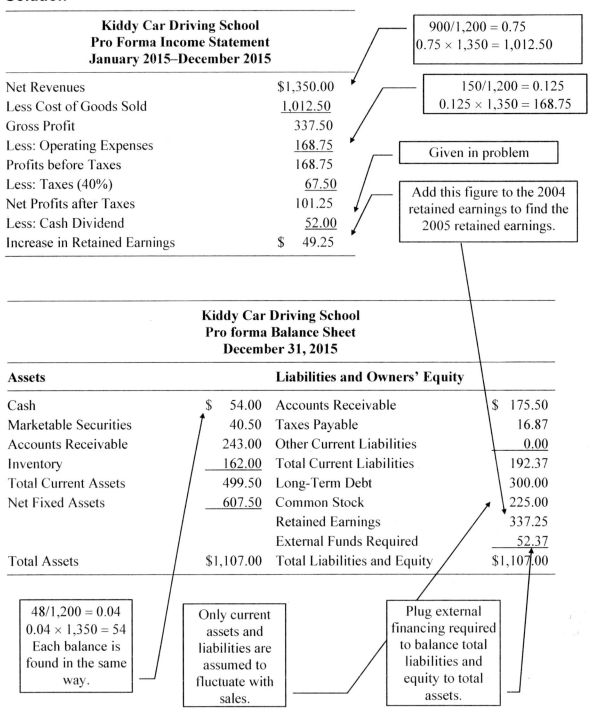

Study Tips

1. Most instructors will emphasize the weaknesses of percentage-of-sales forecasting methods. Recognize you are assuming that the firm's past financial condition is an accurate indicator of its future and that certain variables, such as cash, accounts receivable, and inventories, can be forced to take on certain desired values. The methods discussed in this chapter are only as valid as these assumptions. For established firms, they may be accurate. For newer, more volatile firms, they may not.

2. Cash budgets are projections about uncertain events. As such, even when the assumptions hold true, sales forecasts may be wrong. This is where scenario analysis becomes important. In more sophisticated models, probabilities will be assigned to various outcomes to aid in the selection of the right level of external financing. Still, in the final analysis, the manager's experience, intuition, and risk tolerance will play an important role in determining the borrowing level.

Student Notes

___

___

___

___

___

Sample Exam—Chapter 4

True/False

T F 1. The strategic financial plans are planned long-term financial actions and the anticipated financial impact of those actions.

T F 2. Cash planning involves the preparation of the firm's income statement.

T F 3. The sales forecast, cash budget, and pro forma financial statements are the key outputs of the short-run (operating) financial planning.

T F 4. The excess cash balance is the amount available for investment by the firm if the desired minimum cash balance is less than the period's ending cash.

T F 5. The cash budget gives the financial manager a clear view of the timing of the firm's expected profitability over a given period.

T F 6. A firm's net cash flow is the mathematical difference between the firm's beginning cash and its cash disbursements in each period.

T F 7. The financial planning process begins with short-run plans and budgets that in turn guide the formulation of long-run financial plans.

T F 8. One weakness of the percentage-of-sales method for developing pro forma income statements is that it assumes the past performance of the firm will be an indicator of future performance.

T F 9. The financial manager may cope with uncertainty and make more intelligent short-term financial decisions by preparing several cash budgets, each based on differing assumptions.

T F 10. If the net cash flow is less than the minimum cash balance, financing is required.

T F 11. Due to the no-fixed-cost assumption in the percentage-of-sales method, the use of cost and expense ratios generally tends to understate profits when sales are increasing and overstate profits when sales are decreasing.

T F 12. A positive external funds requirement would indicate that the firm's financing is in excess of its needs and that funds would therefore be available for repaying debt, repurchasing stock, or increasing the dividend to stockholders.

T F 13. The pro forma statements provide the financial manager with the amount of external financing required to support a given level of sales, as well as a basis for analyzing in advance the level of profitability and overall financial performance of the firm in the coming year.

T F 14. One basic weakness of the judgmental approach to developing pro forma balance sheets lies in the assumption that certain variables, such as cash, accounts receivable, and inventories, can be forced to take on certain "desired" values.

T F 15. The "plug" figure that is used as a balancing account in the pro forma balance sheet is retained earnings.

T F 16. When estimating cash flows, depreciation and any other noncash charges are added back to operating profits.

T F 17. The MACRS method of calculating depreciation is used for tax purposes.

Multiple Choice

1. The financial planning process begins with _________ financial plans that in turn guide the formation of _________ plans and budgets.
   a. long-run; operating
   b. strategic; operating
   c. short-run; long-run
   d. long-run; short-run

2. The key aspects of the financial planning process are
   a. investment planning and profit planning.
   b. cash planning and financing.
   c. cash planning and investment planning.
   d. cash planning and profit planning.

3. Pro forma statements are used for
   a. profit planning.
   b. credit analysis.
   c. cash budgeting.
   d. all of the above.

4. Which of the following would be the least likely to utilize pro forma financial statements or a cash budget?
   a. top management
   b. middle management
   c. investors
   d. lenders

5. The __________ is a financial projection of the firm's short-term cash surpluses or shortages.
   a. operating plan
   b. cash budget
   c. strategic plan
   d. pro forma income statement

6. __________ forecast is based on the relationship between the firm's sales and certain economic indicators.
   a. A sales
   b. An internal
   c. A pro forma
   d. An external

7. Key inputs to short-term financial planning are
   a. leverage analysis.
   b. economic forecasts.
   c. sales forecasts and operating and financial data.
   d. financial budgets.

8. The key input to any cash budget is the
   a. sales forecast.
   b. current economic outlook.
   c. operating plan.
   d. pro forma income statement.

9. __________ forecast is based on a buildup of sales forecasts through the firm's own sales channels, adjusted for additional factors such as production capabilities.
   a. An external
   b. A pro forma
   c. An internal
   d. A sales

10. The firm's final sales forecast is usually a function of
    a. external indicators.
    b. both internal and external factors.
    c. internal forecast.
    d. top management's estimates.

11. Of the following, generally the easiest to estimate are
    a. cash sales.
    b. cash disbursements.
    c. cash receipts.
    d. short-term borrowings.

12. A projected excess cash balance for the month may be invested in
    a. marketable securities.
    b. junk bonds.
    c. long-term securities.
    d. secured bonds.

13. The percentage-of-sales method of preparing the pro forma income statement assumes that all costs are
    a. fixed.
    b. independent.
    c. constant.
    d. variable.

14. All of the following accounts are used in the preparation of the cash budget EXCEPT
    a. cash receipts.
    b. depreciation.
    c. cash disbursements.
    d. excess cash.

15. If the firm expects short-term cash surpluses, it can plan
    a. short-term investments.
    b. leverage decisions.
    c. long-term investments.
    d. short-term borrowing.

16. The ________ method of developing a pro forma balance sheet estimates values of certain balance sheet accounts, while others are calculated.
    a. accrual
    b. percentage-of-sales
    c. judgmental
    d. cash

17. __________ generally reflects the anticipated financial impact of long-term actions.
    a. An operating financial plan
    b. A strategic financial plan
    c. A pro forma income statement
    d. A cash budget

18. The primary purpose in preparing a budget is
    a. for profit planning.
    b. for risk analysis.
    c. to estimate sales.
    d. for cash planning.

19. Generally, firms that are subject to a high degree of __________, relatively short production cycles, or both, tend to use shorter planning horizons.
    a. financial planning
    b. profitability
    c. financial uncertainty
    d. operating uncertainty

20. All of the following are key economic indicators EXCEPT
    a. internal sales forecast.
    b. disposable personal income.
    c. GNP.
    d. new housing starts.

21. Kein anticipates take-home pay of $2,750 in each of the next three months, plus a dividend of $75 in the second month. His primary expenses are housing at $556 per month and food, which averages $345 per month. Other expenses are expected to be $1,045 per month. He anticipates spending $2,400 on a new plasma television in the third month. What is his cash surplus (or deficit) in the second month?
    a. $729
    b. $879
    c. $1,705
    d. $1,780

22. Going back to the prior question, what is the cumulative cash surplus (or deficit) over the three months?
    a. $ 87
    b. $237
    c. $2,412
    d. $2,487

23. Which of the following is not correct about personal goals?:
    a. Personal goals should be clearly defined.
    b. Personal goals should have a priority.
    c. Personal goals should be the same throughout one's life.
    d. Personal goals should include a cost estimate.

24. A corporation just purchased a new machine costing $40,000, with a salvage value of $5,000 and installation costs of $3,000. The machine will be depreciated using MACRS with a seven-year recovery period. The total depreciable cost is ______ and the number of years of depreciation expense is ____.
    a. $40,000; seven
    b. $43,000; seven
    c. $43,000; eight
    d. $38,000; eight

25. Which of the following equations is the equation for OCF (operation cash flow)?
    a. OCF = EBIT × (1 – T)
    b. OCF = [EBIT × (1 –T)] + Depreciation
    c. OCF = [EBIT × (1 –T)] – Net fixed asset investment
    d. OCF = [EBIT × (1 –T)] – Net fixed asset investment – Net current asset investment

Essay

1. Discuss the purpose of spending so much time and energy preparing cash budgets and pro forma financial statements. Is this effort well spent even in firms that have an obvious cash surplus?

2. Discuss the weaknesses of the simplified approaches to preparing pro forma financial statements.

3. Explain why depreciation, amortization and depletion are noncash charges.

■ Chapter 4 Answer Sheet

True/False

1. T
2. F
3. F
4. T
5. F
6. F
7. F
8. T
9. T
10. F
11. T
12. F
13. T
14. T
15. F
16. T
17. T

Multiple Choice

1. D
2. D
3. D
4. C
5. B
6. D
7. C
8. A
9. C
10. B
11. C
12. A
13. D
14. B
15. A
16. C
17. B
18. D
19. D
20. A
21. B
22. A
23. C
24. C
25. B

Essay

1. Cash budgets and pro forma statements allow the financial manager to anticipate the need for funds well in advance. By having prior warning of an impending problem, actions can be taken to deal with it. For example, short-term lines of credit can be established. Additional equity can be sold, or the firm can choose a more moderate level of growth. Even in firms that have cash surpluses, cash budgeting is important. By determining how long funds can be invested, the financial manager can take advantage of the higher returns usually available on long-term securities.

2. The basic weakness of the simplified approaches relates to the underlying assumptions. The financial manager assumes that the past is a relevant and accurate predictor of the future. In addition, because the financial manager must make projections, there is the risk that desired rather than real projections will be used.

3. Depreciation, amortization, and depletion are noncash charges because they are expenses that are deducted on the income statement but do not involve an actual outlay of cash. Therefore when measuring cash flow, depreciation must be added back to net income or cash would be understated.

Chapter 5
Time Value of Money

■ Chapter Summary

Without question, this is the most important chapter in the text. Not because the concepts are necessarily more important than those presented in other chapters, though they are fundamental to all of finance, but because most of the rest of the text will use the methods developed here. If you fail to really internalize this material, you will have a very difficult time with the rest of the course. On the other hand, by studying this chapter carefully, you will find that finance is both interesting and understandable. Because of the importance of this material, the study guide has more examples and practice problems than usual. Do them all!

LG 1 *Discuss the role of time value in finance, the use of computational tools, and the basic patterns of cash flow.* The timing of cash flows must be considered because financial managers make decisions affecting cash flows occurring over many years. A dollar received today is worth more than a dollar received tomorrow if for no other reason than because it could be invested. There are several tools available to help with time value calculations, including calculators and built-in functions on spreadsheet programs.

LG 2 *Understand the concepts of future value and present value, their calculation for single amounts, and the relationship between them.* Future values are amounts that will accrue by letting a sum of money earn income over time. The longer a balance is allowed to earn interest, the greater the future value will be. Compound interest is interest that is earned on interest. The more frequently the balance is compounded, the higher the effective interest rate. The concept of present value involves discounting future cash flows back to the present to determine what you would be willing to pay today for the right to receive the future cash flow. Present values are used extensively in business to analyze business opportunities and investments.

LG 3 *Find the future value and the present value of both an ordinary annuity and an annuity due, and find the present value of a perpetuity.* An annuity is a stream of payments where each payment is exactly alike and where the payments occur at exactly the same interval. For example, if you win the lottery, the 20 equal annual payments you receive is an annuity. The payments do not have to occur annually. An annuity can have payments received monthly, weekly, or even daily; they just have to be equally spaced and of equal amounts. Because annuities occur so often in business, there are time-saving techniques for finding their value.

LG 4 *Calculate both the future value and the present value of a mixed stream of cash flows.* A mixed stream of cash flows is one where the amounts to be received vary each period, as opposed to an annuity where the cash flows are equal across periods. There is often a need to find the present value or future value of a stream of cash flows that varies. For example, a business opportunity may involve one stream of cash flows for the first 5 years and a different stream during the next 10 years. When valuing assets, we will use the present value methodologies.

LG 5 *Understand the effect that compounding interest more frequently than annually has on future value and on the effective annual rate of interest.* We frequently need to work with cash flows that are received more frequently than once per year or are paid more frequently than once per year. For example, savings institutions compound interest monthly, daily, or even continuously.

LG 6 *Describe the procedures involved in (1) determining deposits needed to accumulate a future sum, (2) loan amortization, (3) finding interest or growth rates, and (4) finding an unknown number of periods.* There are many common applications of time value of money. For example, to determine how much must be set aside to accumulate some future balance. Suppose that you have determined that you will need $1 million to retire. If you have 20 years until retirement, you would need to know how much must be put into your retirement account each year to reach your retirement goal.

■ Chapter Notes

Time Value of Money

LG 1 Recall from Chapter 1 that the financial manager must make decisions based on cash flows. Because a firm can make positive returns on these cash flows, the financial manager must take into consideration the time value of money. There are two common views in which a financial manager can measure the time value of money: future value and present value.

1. Future value techniques are used to find future values, which are measured at the end of a project's life. The future value technique uses compounding to find the future value of each cash flow at the end of a project's or investment's life and then sums each future value to find the future value of the project.
2. Present value techniques are used to find present values, which are measured at the beginning of a project's life. The present value technique uses discounting to find the present value of each cash flow at time zero and then sums them to find the investment's present value.

The cash flow—both inflows and outflows—of a firm can be described by its general pattern. It can be defined as a single amount, an annuity, or a mixed stream.

Single amount: A lump-sum amount either currently held or expected at some future date. Examples include $1,000 today and $650 to be received at the end of 10 years.

Annuity: A level periodic stream of cash flow. For our purposes, we'll work primarily with *annual* cash flows. Examples include either paying out or receiving $800 at the end of each of the next seven years.

Mixed stream: A stream of cash flow that is *not* an annuity; a stream of unequal periodic cash flows that reflect no particular pattern.

Computational Aids

The calculation of future and present value is aided by the use of calculators and spreadsheets.

There are a variety of inexpensive and very powerful calculators available, and they are highly recommended. The Financial Calculator section of this handbook provides some tips on the use of the more popular models.

Personal computers using spreadsheet programs can also perform TVM calculations.

Future Value and Present Value of a Single Amount

LG 2 The future value of a present amount is found by applying compounded interest over a specified period of time.

The calculation of future values can be simplified in several ways.

1. To avoid compounding interest at each year's end, the following equation can be used to find the future value for a single amount of money, over any number of periods.

$$FV_n = PV \times (1 + r)^n$$

where:

FV_n = the future value at the end of period
PV = initial principal, or present value
r = the annual rate of interest paid
n = the number of periods, typically years

Example

If Tom places $100 into a savings account that earns 6% interest compounded annually, at the end of year 1, he will have $100 in principal and $6 in interest for a total of $106.

$$FV\ at\ end\ of\ year\ 1 = \$100 \times (1 + 0.06) = \$106$$

If Tom leaves the $106 in the account for another year, the principal plus the interest earned in year 1 will be compounded for an additional year. The new amount at the end of year 2 will be $112.36.

$$FV\ at\ end\ of\ year\ 2 = \$106 \times (1 + 0.06) = \$112.36$$

The $112.36 is the future value of $100 compounded annually at 6% for 2 years.

Using the example above, the future value for $100 compounded annually at 6% for 2 years can be found by:

$$FV_n = \$100 \times (1 + 0.06)^2$$
$$FV_n = \$112.36$$

2. Financial calculators are available today at very reasonable prices and can ease solving many types of time value problems. It is best to begin by learning to solve the problems using the equations, then to progress to using the calculator. By using the formulas, you learn to adjust for nonannual compounding and other issues that will surface later in the chapter. These issues do not disappear when you begin using a calculator.

A financial calculator is distinguished by a row of buttons that correspond to the inputs to TVM problems, where:

N = the number of periods
I = the interest rate per period
PV = the present value or the initial deposit
CMT = the payment (this button is used only when there is more than one equal payment)
FV = the future value

These buttons appear either on the face of the calculator or on the calculator's screen.

There are a few common errors students make when using financial calculators. First, be sure that the number of compounding periods is set correctly. The default is usually12 periods per year. If the number of periods is set to 12 periods per year, the calculator divides the interest rate by 12 before performing calculations. Most new calculators are preprogrammed to divide the interest rate by 12. In addition, whenever the batteries are replaced the calculator will revert to dividing by 12. (Check your calculator now and set it to one period per year. To check the number of periods on the TI BA-II plus press the 2nd key and then the I/Y button. Your calculator should display P/Y = 1.00. If not, enter 1 on the keypad and press Enter. For other calculators, refer to your owner's manual.)

Using the previous example, the future value for $100 compounded annually at 6% for 2 years can be found using the calculator by:

$$N = 2$$
$$I = 6$$
$$PV = 100$$
$$PMT = 0$$
$$CPT\ FV = -\$112.36$$

The calculator differentiates inflows from outflows by preceding the outflows with a negative sign. For example, in the problem just demonstrated, the $100 present value (*PV*), because it was keyed as a positive number, is considered an inflow. Therefore, the calculated future value (*FV*) of –112.36 is preceded by a minus sign to show that it is the resulting outflow. Had the $100 present value been keyed in as a negative number (–100), the future value of $112.36 would have been displayed as a positive number (112.36). Simply stated, *the cash flows—present value* (*PV*) *and future value* (*FV*)*—will have opposite signs.*

3. Using the previous example we can use an Excel spreadsheet to solve this problem as follows:

	A	B
1	**Future Value of a Single Amount**	
2	Present Value	100
3	Interest Rate	0.06
4	Number of Years	2
5	Future Value	112.36
Entry in Cell B5 is = *FV*(B3,B4,0, –B2,0)		

Present Value of a Single Amount

The present value of a single amount is found by discounting each year's cash flows back to time zero. The process of discounting is simply the inverse of compounding interest. The application of discounting is concerned with answering the question, "If I am earning *r* percent on my money, what will I be willing to pay today, in year zero, for the opportunity to receive FV_n dollars *n* periods from now?"

The equation for finding the present value of a single amount can be stated as:

$$PV = \frac{FV_n}{(1+r)^n} = FV \times \left[\frac{1}{(1+r)^n}\right]$$

Example

Harry wishes to find the present value of $500 that will be received five years from now. Harry is earning 10 percent on his money. What is the most that Harry is willing to pay today to receive the $500 five years from now?

1. Equation Solution

$$PV = \$500 \times \left[\frac{1}{(1+0.10)^5}\right]$$

$$PV = \$310.46$$

2. Calculator Solution

$$N = 5$$
$$I = 10$$
$$FV = 500$$
$$PMT = 0$$
$$CPT\ PV = \$310.46$$

3. Excel Spreadsheet

	A	B
1	**Present Value of a Single Amount**	
2	Future Value	500
3	Interest Rate	0.1
4	Number of Years	5
5	Present Value	310.46
Entry in Cell B5 is $= -PV$(B3,B4,0,B2,0)		

Future Value and Present Value of an Annuity

LG 3

An annuity is a stream of equally spaced, equal cash flows. The two basic types of annuities are the ordinary annuity and the annuity due. With an ordinary annuity, payments occur at the end of each period, and with an annuity due, payments occur at the beginning of each period.

The future value of an annuity can be found by summing the future value of each year's cash flow. It is important to remember that with an ordinary annuity, payment is made at the end of the period, so in the last period no interest will be received.

You can calculate the future value of an ordinary annuity that pays an annual cash flow equal to CF by using the following equation:

$$FV_n = CF \times \left\{ \frac{\left[(1+r)^n - 1\right]}{r} \right\}$$

The payments for an annuity due are received at the beginning of each period. Because of this, the equation for finding the future value of an annuity must be converted to work for the future value of an annuity due. The FV_n is converted for an annuity due as follows:

$$FV_n \text{ (annuity due)} = FV_n \times (1+r)$$

As can be seen from the equations above, the future value of an annuity due is more than the future of an ordinary annuity because the payments are at the beginning of the period so they earn interest for one more period.

Example

Sally wishes to find the future value of $100 that will be received annually at the end of each year for the next five years, and Sally is earning 10 percent on her money.

1. Equation

$$FV = 100 \times [(1.1)^5 - 1] / 0.1 = 610.51$$

2. Calculator

$$N = 5$$
$$I = 10$$
$$PV = 0$$
$$PMT = 100$$
$$CPT \ \ FV = \$610.51$$

3. Excel Spreadsheet

	A	B
1	**Future Value of an Ordinary Annuity**	
2	Annual Payment	100
3	Annual Rate of Interest, Compounded Annually	0.10
4	Number of Years	5
5	Future Value of an Ordinary Annuity	610.51
Entry in Cell B5 is = *FV*(B3,B4, –B2,0,0)		

Present Value of an Annuity

An annuity is a stream of equal payments that arrive at equal intervals. One approach would be to calculate the present value of each cash flow in the annuity and then add up those present values. Alternatively, the algebraic shortcut for finding the present value of an ordinary annuity that makes an annual payment of CF for n years looks like this

$$PV_n = \left(\frac{CF}{r}\right) \times \left[1 + \frac{1}{(1+r)^n}\right]$$

The payments for an annuity due are received at the beginning of each period. Because of this, the equation for finding the present value of an annuity must be converted to work for the present value of an annuity due. The PV_n is converted for an annuity due as follows:

$$PV_n \text{ (annuity due)} = PV_n \times (1 + r)$$

Example

Sally wishes to find the present value of $100 that will be received annually at the end of each year for the next five years, and Sally is earning 10 percent on her money.

1. Equation

$$PV = (100/0.1) \times [1-(1/(1.1)5)] = 379.08$$

2. Calculator

$$N = 5$$
$$I = 10$$
$$FV = 0$$
$$PMT = 100$$
$$CPT\ PV = -379.08$$

3. Excel Spreadsheet

	A	B
1	**Present Value of an Ordinary Annuity**	
2	Annual Payment	100
3	Annual Rate of Interest, Compounded Annually	0.1
4	Number of Years	5
5	Present Value of an Ordinary Annuity	379.08
Entry in Cell B5 is = *PV*(B3,B4,–B2,0,0)		

Perpetuity

A perpetuity is an annuity that never stops providing its holder with payment dollars at the end of each year. Thus, to find the present value of a stream of payments that goes on forever, simply divide the payment amount by the interest rate.

$$PV_{perpetuity} = CF \div r$$

For example, the present value of a stream of payments of $2 that will go on forever, assuming a 10 percent interest rate, is $2/0.1 = $20.

Future Value and Present Value of Mixed Cash Flow Streams

LG 4

There are two possible types of cash flow streams. A mixed stream of cash flows reflects no pattern year after year. An annuity, as defined earlier, is a stream of equal cash flows.

The present value for a mixed stream of cash flows can be found by finding the present value of each future amount, and then summing them to find the present value of the investment. The future value for a mixed stream of cash flows can be found by finding the future value of each future amount, and then summing them to find the future value of the investment.

Example of Future Value of a Mixed Stream of Cash Flows

Given the following cash flows and an 8 percent interest rate, what will be the future value of the investment?

Year (*n*)	Cash flow	Future Value
1	$700	$881.80
2	$400	$466.56
3	$500	$540.00
4	$300	$300.00
Future value of mixed stream of cash flows		$2,188.36

1. The future value for each year can be found with the equation below:

$$FV_n = PV \times (1 + r)^n$$

or with a financial calculator as demonstrated earlier. Some calculators have a function allowing you to input *all of the cash flows,* specify the interest rate, and directly calculate the future value of the entire cash flow stream.

2. The future value of a mixed stream of cash flow can also be found with an Excel Spreadsheet as illustrated below:

	A	B	C	D	E
1	**Future Value of a Mixed Stream**				
2	Year Cash	Years to Earn	Cash	Interest	FV in
3	Is Received	Interest	Flow	Rate	Year 5
4	1	3	700	8%	$ 881.80
5	2	2	400	8%	$ 466.56
6	3	1	500	8%	$ 540.00
7	4	0	300	8%	$ 300.00
8			Future Value		$2,188.36
9	Entry in Cell E9 is = SUM(E4:E7)				

Example of Present Value of a Mixed Stream of Cash Flows

Texas Oil expects an investment to provide the following cash flows and requires an 8 percent return on its investments. What will be the present value of the investment?

Year (n)	Cash flow	Present Value
1	$700	$648.15
2	$400	$342.94
3	$500	$396.92
4	$300	$220.51
Present value of mixed stream of cash flows		$1,608.51

1. The present value for each year can be found with the equation below:

$$PV = \frac{FV_n}{(1+r)^n} = FV \times \left[\frac{1}{(1+r)^n}\right]$$

or with a financial calculator.

2. Most calculators allow you to calculate the present value of a mixed stream of cash flows directly.

The steps to follow for the financial calculator are illustrated below for this example.

Before using the *CF* button, make sure you clear your calculator by inputting *CF*; 2nd; *CE/C*.

Data and Key Inputs	Display
0; Enter	$CF0 = 0$
Down Arrow; 700; ENTER	$CF1 = 700$
Down Arrow; 1; ENTER	$F01 = 1$
Down Arrow; 400; ENTER	$CF2 = 400$
Down Arrow; 1; ENTER	$F02 = 1$
Down Arrow; 500; ENTER	$CF3 = 500$
Down Arrow; 1; ENTER	$F03 = 1$
Down Arrow; 300; ENTER	$CF4 = 300$
Down Arrow; 1; ENTER	$F04 = 1$
NPV; 8; ENTER	$I = 8$
Down Arrow; *CPT*	$NVP = 1608.51$

3. The present value of a mixed stream of cash flow can also be found with an Excel spreadsheet as illustrated below:

	A	B
1	**Present Value of a Mixed Stream**	
2	Interest rate, %/year	8%
3	Year	Year-End Cash Flow
4	1	700
5	2	400
6	3	500
7	4	300
8	Present Value	1,608.51
9	Entry in Cell B8 is = *NPV*(B2, B4: B7)	

Compounding More Often than Annually

LG 5

Sometimes savings institutions compound interest more frequently than annually. If this is the case, the future value formula can be restated as:

$$FV_n = PV \times \left(1 + \frac{r}{m}\right)^{n \times m}$$

where: m = number of times per year interest is compounded

Referring to the preceding example, Tom's \$100 compounded quarterly at 6 percent for two years will now be worth \$112.65.

$$FV_n = \$100 \times \left(1 + \frac{0.06}{4}\right)^{4 \times 2}$$

$$FV_n = \$112.65$$

Occasionally interest is compounded continuously. The FV used in the continuous compounding formula is found by:

$$FV\ (continuous\ compounding) = e^{r \times n}$$

where: e = the exponential function which has a value of 2.7183

To compare loan costs or investment returns over different compounding periods, nominal and effective interest rates must be distinguished.

Nominal and Effective Annual Rates of Interest

The nominal rate of interest is the contractual or stated interest rate. The effective annual rate reflects the impact of compounding. The following formula converts the nominal rate (r) to its effective rate with m compounding periods per year.

$$EAR = \left(1 + \frac{r}{m}\right)^m - 1$$

Example: What is the effective annual rate if the nominal rate is 12% and deposits are compounded monthly?

Solution:

$$EAR = \left(1 + \frac{0.12}{12}\right)^{12} - 1$$

$$EAR = 0.1268 = 12.68\%$$

Special Applications of Time Value

LG 6

There are four special applications of time value that are introduced in Chapter 5.

1. **Payments to Accumulate a Future Balance:** Often it is necessary to calculate the annual deposits needed to accumulate a certain amount of money n periods from now. The equation that makes it possible to find the necessary deposits is an alteration of the equation used to find the future value of an annuity.

$$FV_n = CF \times \left\{ \frac{\left[(1+r)^n - 1\right]}{r} \right\}$$

To find the necessary deposits, simply solve the equation to determine payments, CF. You could also use a financial calculator.

Example

You want to save the down payment for a house and you can earn 7 percent interest per year. If you need to accumulate $10,000 over the next five years, how much must you save every year?

Using the financial calculator:

$$N = 5$$

$$I = 7$$

$$PV = 0$$

$$FV = 10{,}000$$

$$CPT\ PMT = 1738.91$$

So you would have to save $1,738.91 per year for five years at 7 percent interest to save $10,000.

2. **Loan Amortization:** The term *loan amortization* refers to the determination of the equal annual loan payments necessary to provide a lender with a specified interest return and repay the loan principal over a specified period. To find loan payments, use the present value of an annuity formula and rearrange the terms to solve for *CF*.

$$PV_n = (CF / i) \times [1 - (1/((1 + r)^n)]$$
$$CF = (PV_n \times i) / [1 - (1/((1 + r)^n)]$$

Or use the financial calculator.

Example

What is the annual payment due on a \$10,000 loan repayable with annual payments over 10 years if interest is 10 percent?

a. Equation

$$PMT = (10{,}000 \times 0.1)/[1 - (1/((1.1)10)] = 1627.45$$

b. Calculator

$$N = 10$$
$$I = 10$$
$$PV = 10000$$
$$FV = 0$$
$$CPT\ PMT = 1627.45$$

3. **Interest Growth Rates:** It is often necessary to calculate the annual rate of growth exhibited by a stream of cash flows. For example, you may want to know the annual compounded rate of growth in the dividends paid by a firm. To do this, use only the **first and last** cash flows. Ignore all the cash flows in the middle. Determine how many periods there were when growth could occur, then use financial calculator to solve for the interest rate that will equate the present value with the future value.

Example

Assume that the dividend for Bogus Corp. was \$2.00 in 2003, \$2.25 in 2004, \$2.35 in 2005, \$1.90 in 2006, \$2.00 in 2007, and \$2.50 in 2008. What was the annual compounded growth rate of the dividends?

Remember to ignore all but the first and last cash flow. Thus the *PV* will be \$2.00 and the *FV* will be \$2.50. Then count the *intervals* over which the cash flow could grow. In this example the dividend could grow between 2003 and 2004, between 2004 and 2005, between 2005 and 2006, between 2006 and 2007, and between 2007 and 2008, which is five growth periods, which is also one less than the number of years of data given. (Also, remember either the *PV* or the *FV* must be input as a negative number.)

$$PV = -2$$
$$FV = 2.5$$
$$PMT = 0$$
$$N = 5$$
$$CPT\ I = 4.56\%$$

4. **Finding an Unknown Number of Periods:** Sometimes it is necessary to calculate the number of time periods needed to generate a given cash flow from an initial amount.
Example: How long would it take for \$100 to double to \$200 dollars if you were earning 8 percent?
Using the financial calculator:

$$PV = -100$$
$$FV = 200$$
$$I = 8$$
$$PMT = 0$$
$$CPT\ N = 9.01 \text{ years}$$

Sample Problems and Solutions

1. **Future Value (Compounding)**

$$FV_n = PV(1 + r)^n$$

Example 1. You deposit \$150 in the bank and leave it for five years at an annual rate of 6 percent. Interest is paid annually. How much will be in your account at the end of the fifth year?

a. **Equation Solution**
Using the above equation, $\$150(1.06)^5 = \$200.73$.

b. **Calculator Solution**
Using the calculator,

$$N = 5$$
$$I = 6$$
$$PV = 150$$
$$CPT\ FV = \$200.73$$

Example 2. When you were two years old, your grandmother left you a trust fund with \$10,000 in it that has been earning 12 percent for 20 years. How much is in the fund now (assume annual compounding)?

a. **Equation Solution**
$\$10{,}000(1.12)^{20} = \$96{,}462.93$

b. **Calculator Solution**

$$N = 20$$
$$I = 12$$
$$PV = 10{,}000$$
$$CPT\ FV = \$96{,}462.93$$

Example 3. Now assume that instead of annual compounding, the bank pays interest to you semi-annually (twice per year). Up until now we have assumed that n referred to the number of years. Actually it refers to the number of *periods*. If you have \$100 on deposit, what is it worth at the end of three years if the annual interest rate is 12 percent?

a. **Equation Solution**

Convert the interest rate to the rate that will be earned during one period. Because each period is 1/2 of a year, then the interest rate must be 1/2 of the annual rate = 0.12/2 = 0.06. Next, determine how many periods there are. Because there are two periods per year and three years, there are six periods. The solution is $\$100(1.06)^6 = \$141.85$.

The formula for compounding more than once per year is:

$$FV_n = PV \times \left(1 + \frac{r}{m}\right)^{n \times m}$$

where: m = the number of times per year you compound

b. **Calculator Solution**

$N = 6$

$I = 6$

$PV = 100$

$CPT\ FV = \$141.85$

Example 4. If the bank pays interest quarterly?

a. **Equation Solution**

$\$100(1.03)^{12} = \$142.58$

b. **Calculator Solution**

$N = 12$

$I = 3$

$PV = 100$

$CPT\ FV = \$142.58$

Example 5. If the bank pays interest monthly?

a. **Equation Solution**

$\$100(1.01)^{36} = \$143.08$

Notice that the ending balance increases as the number of compounding periods per year increases.

b. **Calculator Solution**

$N = 36$

$I = 1$

$PV = 100$

$CPT\ FV = \$143.08$

2. **Future Value of an Annuity**

Frequently we are interested in computing how much a series of equal deposits will be worth in the future. For example, say you belong to a retirement program at your work where $1,000 is deposited on your behalf every year. You may want to know how much of a retirement this will provide. You could compute the future value by simply computing the future value of each deposit separately, then adding them all together. The following formula is for computing the future value of an annuity.

$$FV \text{ of Annuity} = PMT\Sigma\,(1 + r)^{t-1}$$

While this method can be extended to any number of periods, it becomes tedious after a while. For this reason we generally solve using the calculator.

Example 1. Compute the future value of $1,000 put into a retirement account for 15 years that earns 10 percent per year.

a. **Calculator Solution**

$PV = 0$
$PMT = 1{,}000$
$N = 15$
$I = 10$
$CPT\ FV = \$31{,}772.48$

Example 2. You make investments of $100 per month for two years. If you earn an average of 12 percent per year, assuming monthly compounding, how much will your investment be worth?

a. **Calculator Solution**

$PV = 0$
$PMT = 100$
$N = 24$
$I = 1$
$CPT\ FV = \$2{,}697.35$

Example 3. Suppose that over the next five years you want to accumulate $45,000 to buy a used Porsche 911. How much must you deposit every year if you can earn 10 percent per year?

a. **Calculator Solution**

$FV = 45{,}000$
$N = 5$
$I = 10$
$CPT\ PMT = \$7{,}370.89$

3. **Present Value**

In earlier work we recognized that because of compound interest income money grows over time. If money grows over time, we would rather have a fixed amount now than that same fixed amount at some time in the future. This is because we can invest the money if we get it now and it will be worth more in the future. But suppose we can't get the money now and instead have to wait until the future before we get paid. That future payment is not worth as much to us as it would have been had it been paid now. This is the concept behind present value. The present value calculation lets us determine how much a sum received in the future is worth today.

Example 1. Compute the present value of $1,000 to be received in one year if interest rates are 10 percent.

a. **Equation Solution**

$PV = FV/(1 + r)$
$PV = \$1{,}000/(1.1) = \$909.09$

b. **Calculator Solution**

$N = 1$
$PMT = 0$
$I = 10$
$FV = 1{,}000$
$CPT\ PV = \$909.09$

Example 2. Compute the present value of $1500 to be received in five years if interest rates are 12 percent.

a. Equation Solution

$PV = \$1,500/(1.12)^5 = \$851.14$

b. Calculator Solution

$N = 5$

$PMT = 0$

$I = 12$

$FV = 1,500$

$CPT\ PV = \$851.14$

Example 3. You get your taxes done at H&R Block and find that you are due a refund of $1,000. Suppose that the refund won't be paid by the IRS for exactly two months. If you feel that you can earn 12 percent on your money, how much does H&R Block have to offer you now for your future payment of $1,000?

a. Equation Solution

$PV = \$1,000/(1.01)^2 = \$980.30$

Because the payment will be received in two months, a monthly rate of interest must be used. Twelve percent per year converts to 1 percent per month. Because two months will pass, the exponent will be 2. The interpretation of the answer is that you will be satisfied if H&R Block offers $980.30 or more for an assignment of your refund.

b. Calculator Solution

$N = 2$

$I = 1$

$FV = 1,000$

$CPT\ PV = \$980.30$

Example 4. Your aunt dies and leaves you $100,000 to be delivered in five years. An attorney offers you $60,000 for an assignment of the trust. If your required rate of return is 12 percent, should you agree to the deal?

a. Equation

$$PV = \$100,000/(1.12)^5 = \$56,742.69$$

If you had $56,742.69 today, it could be invested so that at the end of five years you would have $100,000. If you had the $60,000 that the attorney was offering and invested it, you would have more than $100,000 at the end of five years. Take the offer.

b. Calculator Solution

$N = 5$

$I = 12$

$FV = 100,000$

$CPT\ PV = \$56,742.69$

Example 5. If multiple cash flows that arrive at different times are to be evaluated, compute the *PV* of each one separately and then add them together. For example, suppose you win a scholarship that pays $2,000 at the end of the first year, $2,500 at the end of the second year, and $3,000 at the end of the third year. What is the value of this scholarship if your required rate of return is 8 percent?

a. Equation Solution

$PV_{\$2,000} = \$2,000/(1.08) = \$1,851.85$

$PV_{\$2,500} = \$2,500/(1.08)^2 = \$2,143.35$

$PV_{\$3,000} = \$3,000/(1.08)^3 = \$2,381.50$

$PV_{all} = \$1,851.85 + \$2,143.35 + \$2,381.50 = \$6,376.70$

b. Calculator Solution

$PV = \$1,851.85$

$PV = \$2,143.35$

$PV = \$2,381.50$

$PV_{all} = \$6,376.70$

4. Present Value of an Annuity

An annuity is a stream of equal payments. Note that these payments do not have to be received annually. We can solve this using the formula below:

$$PV \text{ of Annuity} = PMT/(\Sigma(1 + r)^{t-1})$$

Or using a financial calculator.

Example 1. Compute the *PV* of a retirement benefit that pays \$1,500 per year for 20 years if your required rate of return is 12 percent.

a. Calculator Solution

$PMT = 1,500$

$N = 20$

$I = 12$

$CPT\ PV = \$11,204.17$

Example 2. A more difficult problem is where there is a delay before the annuity begins. For example, suppose the above 20-year annuity of \$1,500 did not begin until the end of the fifth year. To solve problems like this, first find the *PVA*. Then, treat that *PV* at time 5 as if it were a lump sum, and find its *PV* at time 0. Find the *PV* if the payment stream in part **c** did not begin for five years.

a. Calculator Solution

$N = 4,$

$PMT = 0$

$I = 12$

$FV = 11,204.17$

$CPT\ PV = \$7,120.45$

5. More on Annuities

The area of working with the time value of money that causes the greatest problems is annuities. Find the present value of the annuities in the following examples, using the timeline of payments and assuming a 10 percent rate of return.

Example 1.

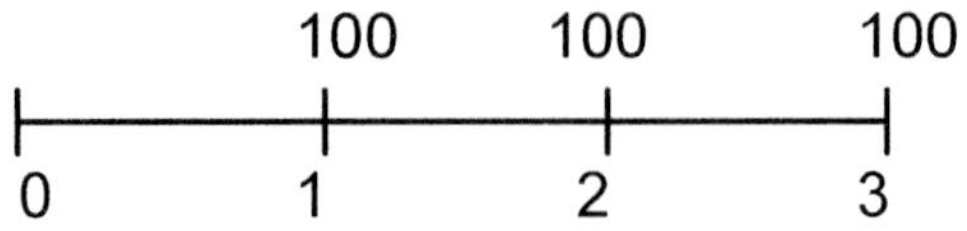

a. Calculator Solution

$PMT = 100$

$FV = 0$

$N = 3$

$I = 10$

$CPT\ PV = \$248.69$

Example 2.

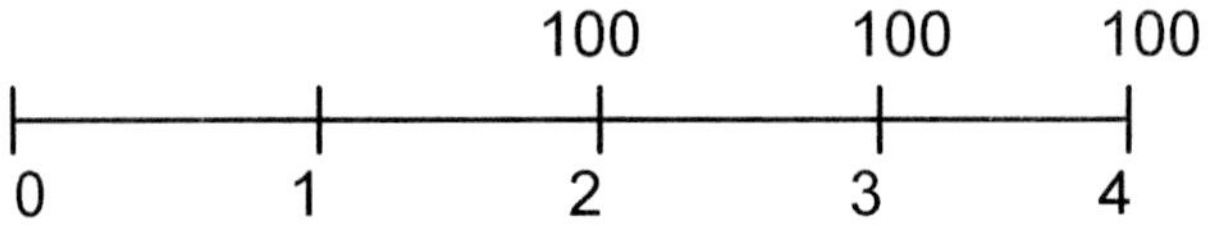

a. Calculator Solution

Step 1:

$N = 3$

$I = 10$

$FV = 0$

$PMT = 100$

$CPT\ PV = \$248.69$

Step 2: This gives us the present value at time 1, so need to discount the value back one more period.

$N = 1$

$I = 10$

$FV = 248.69$

$PMT = 0$

$CPT\ PV = 226.08$

Example 3.

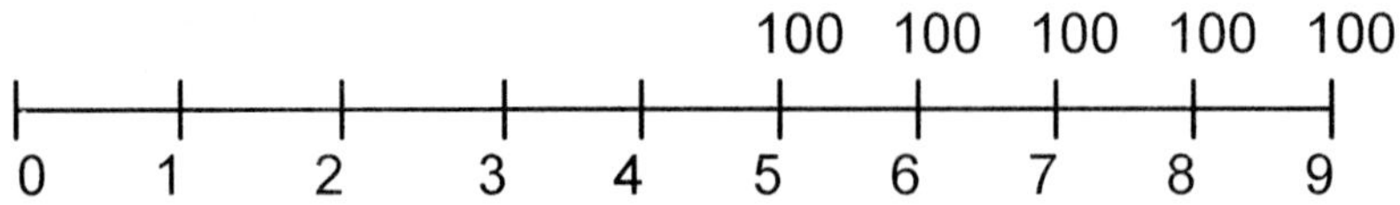

a. Calculator Solution

$PMT = 100, FV = 0$	$N = 4, PMT = 0$
$N = 5$	$I = 10$
$I = 10$	$FV = 379.08$
$CPT\ PV = \$379.08$	$CPT\ PV = \$258.92$

Example 4.

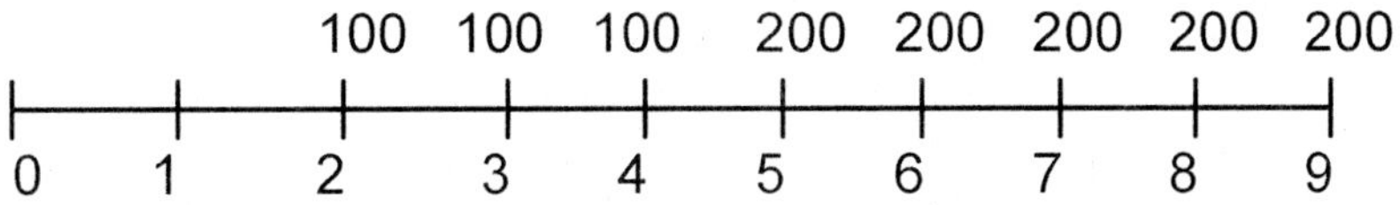

a. Calculator Solution

$PMT = 200$	$PMT = 100$	$PMT = 0$
$N = 5$	$N = 3$	$N = 1$
$I = 10$	$I = 10$	$I = 10$
$FV = 0$	$FV = 758.16$	$FV = 818.30$
$CPT\ PV =$ \$758.16	$CPT\ PV =$ \$818.30	$CPT\ PV =$ \$743.91

6. Loan Amortization

One application of the present value of an annuity formula is to compute payments or loan terms. When a bank makes a loan it must be indifferent to whether it has the money it loans out or it receives the payment stream that the borrower promises to pay. This means that the *PV* of the payment stream must equal the loan amount. To solve a loan amortization problem, fit the parts of the equation that are given.

Example 1.
What is the payment to amortize a \$10,000 loan over 10 years if the interest rate is 12 percent?

a. **Calculator Solution**

$PV = 10{,}000$, FV=0
$N = 10$
$I = 12$
$CPT\ PMT =$ \$1,769.84

7. Growth Rates

To compute the growth rate, use the financial calculator and enter the first year of data as the *PV* (enter as a negative number), enter the last year of data as the *FV*, enter one less than the number of years of data as *N*, and enter 0 as the *PMT* and solve for *I*, the growth rate.

Example 1. You have earnings in year 1 of \$20, year 2 of \$18, year 3 of \$25, year 4 of \$30, and year 5 of \$40. What is the average annual compound earnings growth rate?

Solution

$PV = -20$
$FV = 40$
$N = 4$
$PMT = 0$
$CPT\ I = 18.92\%$

8. Annuity Due

An annuity due is a cash flow that begins at the beginning of each period rather than at the end of each period. For example, rent is usually paid in advance, so it would be an annuity due. To convert from an ordinary annuity, multiply by $(1 + r)$, or when using financial calculators, it is important to set the calculator correctly as to when payments are received (i.e., end vs. beginning mode).

Example. \$500 rent payments are to be made at the beginning of months 1–12. What is the PV of the rental payment stream assuming a cost of capital of 12 percent?

a. Calculator Solution

Before using your calculator to find the present value of an annuity due, depending on the specifics of your calculator, you must either switch it to BEGIN mode or use the DUE key. (*Note:* Because we nearly always assume end-of-period cash flows, *be sure to switch your calculator back to END mode when you have completed your annuity-due calculations.*)

$N = 12$, FV=0

$I = 1$

$PMT = 500$

$CPT\ PV = \$5{,}683.81$

9. Continuous Compounding

So far we have only dealt with discrete compounding intervals, that is, annual, semiannual, quarterly, etc. It is possible to continue compounding beyond this. We could obviously go to daily compounding, then hourly, then by each minute and second. Using calculus, this can be extended into infinity. The resulting compounding equations for continuous compounding are:

$FV_n = PV(e^{r \times n})$ and

$PV_n = FV/(e^{r \times n}) = FV\,(e^{-r \times n})$

where $e = 2.71818182846$

Example

Compute the future value of \$100 invested for 1 year with a compounded 10 percent return.

Solution

$FV = \$100(2.7181810.1(1)) = \$110.5167$

Note that continuous compounding results in a higher ending balance than is possible with any other compounding frequency.

10. Effective Rate

As the number of compounding periods increases, the effective rate also increases. This means that you may be just as well off with a lower rate that is compounded more often than with a simple interest rate. To compare nominal rates to compounded rates, an effective rate must be computed. The formula for computing the effective rate is:

$$EAR = \left(1 + \frac{r}{m}\right)^m - 1$$

where r = annual percentage rate and m = the number of compounding periods per year.

Example 1

What is the effective interest rate if a 12 percent annual rate is compounded 12 times per year?

Solution

Effective Rate = $(1 + (0.12/12))^{12} - 1 = (1.01)^{12} - 1 = 0.1268$

Example 2

Would you rather receive 8.5 percent compounded annually or 8 percent compounded quarterly?

Solution:

Effective Rate = (1.02)4 –1 = 0.0824.

You would rather have the 8.5 percent rate that is compounded annually.

10. Solving for an Unknown Number of Periods

Sometimes it is necessary to calculate the number of time periods needed to generate a given amount of cash flow from an initial amount, and sometimes it is necessary to find the unknown life of an annuity. In these cases you will be solving for the number of periods, n. You will need to use your calculator or a spreadsheet.

Example 1

How long would it take for $100 to double if you can earn 10 percent annually?

a. Calculator solution:

PV= –100

FV=200

PMT=0

I=10

CPT N = 7.27 years

(Note = Either PV or FV must be entered as a negative number, the PMT must be zero because you are not adding any addition funds and the answer N is in years because you are using an annual rate.)

b. Spreadsheet solution:

	A	B
1	**Years for a Present Value to grow to a specified Future Value**	
2	Present Value	100
3	Annual Rate of Interest, compounded annually	0.1
4	Future Value	5
5	Number of Years	379.08
	Entry in Cell B5 is = N*PER*(B3,0,B2,–B4,0,0) The minus sign appears before B4 because the future value is treated as an outflow.	

Example 2:

You can borrow $250,000 at a 12 percent interest rate compounded monthly with equal end of month payments of $2750. How long will it take you to fully repay the loan?

a. Calculator Solution

PV=250,000

FV=0

I=12/12=1

PMT= -2750

CPT N = 64.99 months

(Note: The interest rate is divided by 12 because it is a monthly rate, the PMT is a negative number because it is considered an outflow of cash, and the answer N is in months because a monthly rate was used.)

b. Spreadsheet Solution

	A	B
1	**Months to pay off loan**	
2	Monthly Payment	2750
3	Annual Rate of Interest, compounded monthly	0.1
4	Present Value (loan amount)	250000
5	Number of months to pay off loan	64.99
	Entry in Cell B5 is = N*PER*(B3,–B2,B4) The minus sign appears before B2 because the payments are treated as an outflow.	

■ Quick Drill Self Test

These are quick and easy problems to practice on. The answers follow.

1. What is the future value of a 10-year ordinary annuity of $15,000 per year assuming a discount rate of 10 percent per year?

2. What is the future value of a two-year ordinary annuity of $500 per month, assuming an annual interest rate of 12 percent per year?

3. What is the future value of a $1,000 deposit made today that will be left on deposit for five years earning an annual rate of 10 percent?

4. What is the present value of an ordinary annuity of $1,500 per year that will be received for five years assuming an interest rate of 12 percent?

5. What is the present value of an annuity due of $1,500 per year that will be received for five years, assuming an interest rate of 12 percent?

6. What is the present value of a lump sum of $300 that will be paid at the end of the fifth year assuming that the interest rate is 10 percent?

7. What is the future value at the end of year 3 of the following cash flow stream assuming a 10 percent interest rate? At the end of year 1, $100 will be received, $200 will be received at the end of year 2, and $300 will be received at the end of year 3.

8. What will be the present value of the above cash flow?

9. If the present value of a five-year ordinary annuity is $500, assuming that the discount rate is 10 percent, what are the annual payments?

10. What is the future value after five years of $1,000 that is placed in a savings account today that pays interest quarterly, assuming an annual interest rate of 8 percent?

11. What is the effective annual interest rate for an account that pays interest quarterly at the annual rate of 8 percent?

12. In today's dollars, you need $50,000 to retire. If you will retire in 30 years and if inflation will average 4 percent per year between now and when you retire, how much will you need during the first year of your retirement?

13. What will be the monthly payment on a loan of $15,000 that is amortized over two years at an annual interest rate of 12 percent?

14. What is the interest rate on a loan of $20,000 due in five years and requiring an annual payment of $5,686.29?

15. What is the present value of a perpetuity that pays $2 per quarter if the discount rate is 12 percent per year?

Answers to Quick Drill Self Test

1. $239,061.37
2. $13,486.73
3. $1,610.51
4. $5,407.16
5. $6,056.02
6. $186.28
7. $641
8. $481.59
9. $131.90
10. $1,485.95
11. 8.24 percent
12. $162,169.88
13. $706.10
14. 13 percent
15. $66.67

■ Study Tips

One of the most difficult parts of solving time value of money problems is figuring out which equation to use. When you are doing your homework, this is not too hard because most of the problems are grouped together by problem type. The difficulty arises when you go into your exam and the problems are all mixed up and not labeled. The following *rules of thumb* can be helpful for deciding how to go about finding a solution.

1. Use the *PV* to analyze investments.
2. Use *PV* to compute loan terms.
3. Use *PV* to compute the nest egg needed to make a series of future payments (for example, how much must be saved before you can retire).
4. Use *PV* to compute how much you will pay for something (valuation).
5. Use *FV* to compute a future balance.
6. Use *FV* to compute the payment needed to achieve a future balance.

One last time, this **chapter is critical to your success** in the rest of this course. A great deal of what follows in later chapters will use the methodology developed here. Attempt all of the sample problems. Mark the ones you miss, and come back and try them again later. Keep working these problems until you thoroughly understand how they are done.

■ Student Notes

____

____

____

____

____

■ Sample Exam—Chapter 5

True/False

T F 1. Financial managers make their decisions at time zero, so they tend to use present value techniques.

T F 2. Future value techniques use future values, which are usually measured at the end of a product's life.

T F 3. To find the future value of a single amount of money, the financial manager uses discounting.

T F 4. Quarterly compounding will result in a lower amount of money accumulated than continuous compounding.

T F 5. In personal finance, the nominal rate of interest is used synonymously with the annual percentage rate (APR).

T F 6. The future value of an ordinary annuity is always greater than the future value of an identical annuity due.

T F 7. The effective rate of interest differs from the nominal rate in that it reflects the impact of compounding frequencies.

T F 8. An annuity due is an annuity for which the payments occur at the end of each period.

T F 9. The amortization of a loan involves creating an annuity out of a present amount.

T F 10. A dollar that is received in year 2 will be more valuable than a dollar received in year 1, assuming a constant rate of interest.

T F 11. The future value indicates the amount of money that would have to be invested today at a given interest rate over a specified period to equal some future amount.

T F 12. A mixed stream of cash flows consists of cash flows with no particular pattern.

T F 13. An annuity is a stream of equal periodic cash flows, over a specified time period.

T F 14. To provide a correct answer, financial calculators and electronic spreadsheets require that a calculations relevant cash flows be entered accurately as either cash inflows (positive values) or cash outflows (negative values).

Multiple Choice

1. The rate of interest that is actually paid or earned is called the _______ interest rate.
   a. real
   b. effective
   c. nominal
   d. stated

2. The future value of $500 received today and deposited at 10 percent for 5 years is
   a. $500.
   b. $310.50.
   c. $805.25.
   d. $788.65.

3. At an 8 percent rate of interest, which will result in the greatest amount of money accumulated?
   a. annual compounding
   b. quarterly compounding
   c. semiannual compounding
   d. continuous compounding

4. The present value of $800 received at the end of year 6, assuming an opportunity cost of 7 percent, is
   a. $533.07.
   b. $648.76.
   c. $1,200.80.
   d. $747.66.

5. _________ is an annuity that never stops providing its holder with payment dollars at the end of each year.
   a. An annuity due
   b. A warrant
   c. A perpetuity
   d. An ordinary annuity

6. Occasionally the financial manager will use _________ to roughly estimate how long a given sum must earn at a given annual rate to double that amount.
   a. time-to-double tables
   b. the rule of 72
   c. derivatives
   d. nominal interest rates

7. The future value of an ordinary annuity with $200 annual deposits into an account paying 6 percent interest over the next 10 years will be
   a. $2,120.
   b. $2,794.37.
   c. $1,987.56.
   d. $2,636.16.

8. The present value of a $10,000 perpetuity at an 8 percent discount rate is
   a. $125,000.
   b. $9,259.
   c. $10,000.
   d. $108,000.

9. The present value of $300 received at the end of year 1, $400 received at the end of year 2, and $200 received at the end of year 3, assuming an opportunity cost of 12 percent, is
   a. $803.57.
   b. $729.10.
   c. $1,008.
   d. $648.76.

10. ________ refers to the amount of money on which interest is paid.
    a. Capital
    b. APR
    c. Loan base
    d. Principal

11. Find the future value at the end of year 3 of the following stream of cash flows received at the end of each year. Assume that the firm can earn 9 percent on its investments.

Year	Amount
1	$600
2	800
3	900

    a. $1,918.77.
    b. $2,484.86.
    c. $2,708.40.
    d. $2,507.00.

12. The future value of an annuity due with $400 annual deposits into an account paying 7 percent interest over the next eight years is
    a. $4,391.20.
    b. $3,200.75.
    c. $3,976.36.
    d. $2,800.

13. The future value of $3,000 deposited at 11 percent compounded quarterly for each of the next six years is
    a. $5,211.
    b. $5,611.
    c. $4,976.
    d. $5,753.

14. A young couple wishes to accumulate $35,000 at the end of four years so that they may make a down payment on a house. What should their equal end-of-year deposits be to accumulate the $35,000, assuming a 6 percent rate of interest?
    a. $7,718.
    b. $8,001.
    c. $6,915.
    d. $8,765.

15. The effective interest rate for a savings account that is compounded quarterly at 8 percent is
    a. 8.24 percent.
    b. 8.04 percent.
    c. 8.76 percent.
    d. 9 percent.

16. What will the equal annual end-of-year payments need to be to fully amortize a $25,000, 12 percent loan over a 5-year period?
    a. $3,953.
    b. $6,354.
    c. $4,283.
    d. $6,935.

17. The present value of $10,000 received in year 20, assuming an opportunity cost of 14 percent, is
    a. $480.
    b. $3,280.
    c. $728.
    d. $1,250.

18. The term __________ indicates that the amount earned on a given deposit has become part of the principal at the end of a specified period.
    a. discounting
    b. compounded interest
    c. amortization
    d. future value

19. The future value of an ordinary annuity of $3,500 each year for 25 years, deposited at 9 percent, is
    a. $95,375.
    b. $108,765.24.
    c. $326,486.76.
    d. $296,453.14.

20. Time value of money techniques can be applied to:
    a. lump-sum amounts.
    b. cash flows streams of varying amounts.
    c. streams of similar cash flows.
    d. all of the above.

22. In their professional life, __________ managers need to understand time-value-of-money concepts in order to estimate accurately the cost of investment in new equipment, processes, and inventory.

    a. accounting
    b. operations
    c. marketing
    d. sales

23. Bob Jacobs has been offered either $3,370,000 today or $470,000 per year for 10 years. Bob's opportunity cost is 6 percent compounded annually. Which sum should Bob choose and ***why***?

    a. Bob should choose the $470,000 per year because the present value of this annual payment is $3,459,241, which is greater than $3,370,000.
    b. Bob should choose the $3,370,000 today because its future value of $6,035,157 is greater than the present value of the $470,000, which is $3,459,241.
    c. Bob should choose the $470,000 per year because the future value of this annual payment is $6,194,974, which is greater than the $3,370,000 being offered today.
    d. Bob should choose the $3,370,000 because it is greater than the $470,000 per year.

24. How long will it take $1,000 to triple if you are earning 15 percent annually?

    a. 5.78 years
    b. 4.28 years
    c. 2.66 years
    d. 7.86 years

Essay

1. Explain why the future value of an annuity due is always greater than the future value of an ordinary annuity with the same rate of return and the same amount of periods.

2. Why do financial managers tend to evaluate investment opportunities with present value techniques?

■ Chapter 5 Answer Sheet

True/False

1. T
2. T
3. F
4. T
5. F
6. F
7. T
8. F
9. T
10. F
11. F
12. T
13. T
14. T

Multiple Choice

1. B
2. C $FV = \$500 \times (1.1)^5$
   $FV = \$500 \times 1.61051 = \$805.25$
3. D
4. A $PV = \$800 \times (1.07)^{-6}$
   $PV = \$800 \times 0.66634 = \$533.07$
5. C
6. B
7. D $PMT = 200,\ I = 6,\ N = 10,\ PV = 0\ CPT\ FV$
   $FV = \$2{,}636.16$
8. A $PV = \frac{1}{0.08} = 12.5$
   $PV = \$10{,}000 \times 12.5 = \$125{,}000$
9. B $PV_1 = \$300 \times (1.12)^{-1} = \$267.86$
   $PV_2 = \$400 \times (1.12)^{-2} = \$318.88$
   $PV_3 = \$200 \times (1.12)^{-3} = \$142.36$
   $PV = \$267.86 + \$318.88 + \$142.36 = \$729.10$
10. D
11. B $FV_3 = \$600 \times (1.09)^2 = \$712.86$
    $FV_2 = \$800 \times (1.09) = \$872$
    $FV_1 = \$900$
    $FV = \$712.86 + \$872 + \$900 = \$2{,}484.86$
12. A $PMT = 400,\ I = 7,\ N = 8,\ PV = 0\ CPT\ FV = 4103.92$
    $FVA_{due} = \$400 \times 1.07 = \$4{,}391.20$
13. D $FV = \$3{,}000 \times \left(1 + \frac{0.11}{4}\right)^{4\times 6}$
    $FV = \$3{,}000 \times 1.91763 = \$5{,}753$
14. B $PV = 0,\ I = 6,\ N = 4,\ FV = 35000,\ CPT\ PMT$
    $PMT = 8001$
15. A $k_{eff} = \left(1 + \frac{0.08}{4}\right)^4 - 1$
    $k_{eff} = 8.24\%$
16. D $PV = 25000,\ I = 12,\ N = 5,\ FV = 0\ CPT\ PMT$
    $PMT = 6935$
17. C $PV = \$10{,}000 \times (1.14)^{-20}$
    $PV = \$10{,}000 \times 0.07276 = \$728$
18. B

19. D $PMT = 3{,}500$, $I = 9$, $N = 25$, $PV = 0$, $CPT\ FV$

$FV =$ \$296,453.14

20. D

21. B

22. B

23. A $PMT = 470{,}000$, $I = 6$, $FV = 0$, $N = 10$, $CPT\ PV = 3{,}459{,}241 > 3{,}370{,}000$

24. D $PV = -1000$, $FV = 3000$, $PMT = 0$, $I = 15$, $CPT\ N = 7.86$ years

Essay

1. Because an annuity due's cash flows are received at the beginning of the period, unlike the cash flows of an ordinary annuity, which occur at the end of the period, they have more time to compound. The future value of a dollar received in year 0 will always have a higher future value than a dollar received in year 1.

2. Time value of money techniques are used by financial managers to assess the value of future cash flows. When evaluating the alternatives associated with these future cash flows, financial managers tend to rely on present value techniques because they are operating in time zero. When evaluating projects or investments, financial managers need to answer the question, "What are the future cash flows associated with this opportunity worth to me in present-day dollars?"

Chapter 6
Interest Rates and Bond Valuation

■ Chapter Summary

All assets are valued the same way. Once you have learned the idea behind valuation, you can extend it to the valuation of anything. To value an asset, simply find the present value of any cash flows that will accrue to the holder of the asset. For example, to value a business, find the present value of the cash flows the firm will generate. To find the value of a building, find the present value of the net cash flows the building will generate. The valuation of stocks and bonds are the focus of this chapter. This is a good opportunity to practice your time value of money skills.

LG 1 *Describe the interest rate fundamentals, the term structure of interest rates, and risk premiums.* Interest rates and required returns represent the costs of using various forms of financing. The real rate of interest is the rate that would exist in the absence of risk and inflation. The nominal rate differs from the real rate because it includes inflation and risk premiums. The term structure of interest rates refers to the way yields change depending on how long a security has until it matures.

LG 2 *Review the legal aspects of bond financing and bond cost.* There are certain legal arrangements that help protect bond investors. These are contained in the bond indenture, which includes various provisions of the bond and the bond covenants.

LG 3 *Discuss the general features, yields, prices, ratings, popular types, and international issues of corporate bonds.* There are three features often included in bonds. These include a conversion feature, a call feature, and stock purchase rights. Bonds are rated by several agencies to reflect their risk of default. The ratings range from AAA to D. Eurobonds and foreign bonds enable established creditworthy companies and governments to borrow large amounts internationally.

LG 4 *Understand the key inputs and basic model used in the valuation process.* Every business asset that has value will generate cash flows by either providing profits or by reducing costs. In a later chapter we will focus on how these cash flows are computed. For now, be aware that it is these cash flows that make a business asset valuable. The required return, used as the discount rate when valuing an asset, reflects the riskiness of the cash flows. If you know the cash flows, the timing of the cash flows, and the required return, then you can value any asset.

LG 5 *Apply the basic valuation model to bonds, and describe the impact of required return and time to maturity on bond values.* Regardless of whether you are valuing bonds, stock, or some other asset, you always find the present value of the cash flows. Models will be presented in this chapter for valuing bonds, but keep in focus that models are just applications of the present value techniques you learned in Chapter 5.

LG 6 *Explain yield to maturity (YTM), its calculation, and the procedure used to value bonds that pay interest semiannually.* The yield to maturity is the return an investor will receive if the bond is held until it matures. It is the return that sets the current market value of the bond equal to the present value of the coupon payments and the maturity price. The YTM that is computed for semiannual payments will have to be multiplied by two to get the annual yield.

■ Chapter Notes

Interest Rate Fundamentals

LG 1 When funds are lent, the cost of borrowing the funds is the interest rate.

When funds are obtained by selling ownership interest, as in the sale of stock, the cost to the issuer of the funds is commonly called the required return.

The real rate of interest is the rate that has not been adjusted for inflation, liquidity preferences, or risk. The real rate of interest is the most basic cost of money.

The nominal rate of interest is the actual rate of interest charged by the supplier of funds and paid by the demander. The nominal rate of interest, r, consists of the real rate of interest, r^*, plus an inflation premium, IP, and a risk premium, RP, that covers such things as default risk and contractual provisions.

Nominal Rate of Interest

$$r = r^* + IP + RP$$

where:

r = nominal rate of interest

r^* = real rate of interest

IP = inflation premium

RP = risk premium

The risk free-rate of interest, R_F, is the required return on a risk-free asset. The risk-free rate consists of the real rate of interest plus an inflation premium. The rate on the three-month U.S. Treasury bill (T-bill) is considered to be the risk-free rate of return.

Risk-Free Rate of Return

$$R_F = r^* + IP$$

where:

R_F = risk-free rate of interest

Term Structure of Interest Rates

The term "structure of interest" relates the interest rate to the yield to maturity.

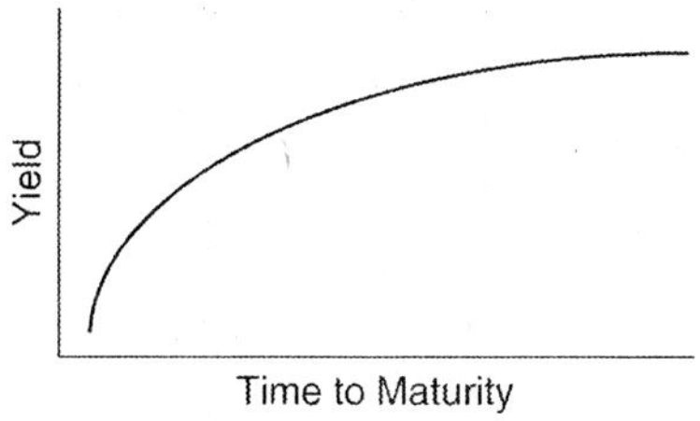

The yield curve is a graph of the term structure of interest rates. It is a graphical representation of the yield to maturity of a security (*y*-axis) and the time to maturity (*x*-axis). The yield curve can take on three different shapes: upward sloping, flat, or downward sloping.

There are three generally cited theories that help to explain the shape of the yield curve.

1. *The expectations hypothesis* reflects investor's expectations about future interest rates. The hypothesis is based on inflationary expectations. Expectations of higher future rates of inflation will result in higher long-term interest rates. The opposite is true for expected lower rates of inflation. If investors believe that inflation rates will be lower in the future, then long-term interest rates will be lower than short-term rates.

2. *Liquidity preference theory* suggests that investors require a premium for tying up funds for long periods of time. This premium is due to the reduced liquidity of long-term funds. To avoid having to repeatedly renew short-term debt, borrowers are willing to pay a premium to obtain long-term funds. This helps to explain why the normal yield curve is upward sloping.

3. *Market segmentation theory* is based on the law of supply and demand. The theory suggests that the market for loans is segmented by the length of maturity for different funds and that the supply and demand for loans within each segment determine the interest rate.

Risk

A risk premium is attached to the risk-free rate to cover such things as default risk, maturity risk, liquidity risk, contractual provisions, and tax risk. In general, securities with the highest risk premiums are considered to have a high rate of default, have a long maturity, contain unfavorable contractual provisions, or not be tax exempt.

There is a risk-return tradeoff, in that investors must be compensated for accepting greater risk with the expectation of greater returns. The higher the risk, the higher the expected return.

Legal Aspects

LG 2

Certain legal arrangements are required to protect purchasers of bonds; bondholders are protected primarily through the indenture and the trustee. The bond indenture is a legal document that specifies both the rights of the bondholders and the duties of the issuing corporation. Included in the indenture are descriptions of the amount and timing of all interest and principal payments, various standard and restrictive provisions, and, frequently, sinking fund requirements.

The standard provisions require that the borrower:

1. maintain satisfactory accounting records
2. periodically supply audited financial statements
3. pay taxes and other liabilities when due
4. maintain all facilities in good working order

Bond indentures also contain the restrictive provisions that keep managers from increasing the risk of the issue. These include requirements that the firm maintain minimum levels of liquidity, limit the sale of accounts receivable to generate cash, constrain subsequent borrowing, and limit the payment of dividends.

The major factors affecting the cost of bond financing, which is the rate of interest paid by the bond issuer, are the bond's maturity, the size of the offering, the issuer's risk, and the basic cost of money.

The current yield is the annual interest payment divided by the current price.

LG 3

There are three features of bonds that are frequently found.

1. A conversion feature allows the holder to convert the bond into stock. This is an option that bondholders will only exercise when the stock price has risen to a point where it is profitable to do so. The conversion feature allows bondholders to share in the good fortunes of the firm.
2. A call feature lets the firm force bondholders to sell the bond back to the firm at the firm's option. This lets the firm retire debt if it is no longer needed or to refinance debt if interest rates fall.
3. Stock purchase warrants are "sweeteners" that give holders the right to purchase shares of common stock at a specified price over a certain period of time.

Bond rating helps investors interpret the risk of a bond issue. Moody's and Standard & Poor's rate bonds from AAA to D, depending on the likelihood of default. The lower the rating, the higher the interest rate the bond must pay to attract investors.

Eurobonds and foreign bonds enable established creditworthy companies and governments to borrow large amounts internationally. A Eurobond is issued by an international borrower and sold to investors in countries with currencies other than the currency in which the bond is denominated. A foreign bond is issued in a host country's financial market, in the host country's currency, by a foreign borrower.

Valuation Fundamentals

Valuation is the process that incorporates the time value of money with the concept of risk and return to determine the worth of an asset.

Key Inputs

LG 4

The value of any asset depends on the cash flows it is expected to provide over time.

In addition to making cash flow estimates, the timing of the cash flows must be known. In combination with one another, the cash flow and its timing fully define the return expected from the asset.

The level of risk associated with a cash flow can affect its value. In general, the greater the risk of the cash flows, the lower its value. Therefore, in the valuation process, the higher the risk, the greater the required return (discount rate).

The Basic Valuation Model

The value of *any* asset is the present value of all future cash flows it is expected to provide over the relevant time period.

The value of an asset based on its cash flows is calculated using the following formula.

$$V_0 = \frac{CF_1}{(1+r)^1} + \frac{CF_2}{(1+r)^2} + \cdots + \frac{CF_n}{(1+r)^n}$$

where:

V_0 = value of the asset at time zero

CF_t = cash flow expected at the end of year t

r = appropriate required return (discount rate)

n = relevant time period

Bond Valuation

LG 5

Bonds are long-term debt instruments used by businesses and governments to raise large sums of money. The value of a bond is the present value of the contractual payments its issuers are obligated to make from the current time until it matures. The interest rate the holder of the bond will receive is called the coupon rate. It does not change while the bond is outstanding. Most corporate bonds have a maturity, or par value, of $1,000. This is the amount the firm pays to the holder when the bond matures. The discount rate used to value bonds is rate of interest bonds with similar risk are yielding in the market at the time the bond is being valued. This rate *will* change over time.

The following equation is used to find the value of a bond making annual coupon payments.

$$B_0 = I \times \left[\sum_{t=1}^{n} \frac{1}{(1+r_d)t} \right] + M \times \left[\frac{1}{(1+r_d)^n} \right]$$

where:

B_0 = value of bond at time zero

I = annual interest in dollars

r_d = discount rate (required return)

n = years to maturity

M = dollar par value

Bond Value Behavior

There are certain external forces that constantly change the value of a bond in the marketplace. Because these forces are not controlled by bond issuers or investors, it is important to understand the impact required return and time to maturity have on bond value.

When the required return on a bond differs from the bond's coupon interest rate, the bond's value will differ from its par value. Increases in the basic cost of long-term funds or in risk will raise the required return. Alternatively, decreases in the basic cost of long-term funds or in risk will lower the required return. When the required return is greater than the coupon interest rate, the bond value will be less than its par value. In this case, the bond will be selling at a discount. When the required return is less than the coupon, the bond value will be greater than par value. In this case, the bond will be selling at a premium. In the following graph, the bond has a coupon interest rate of 10 percent. Notice that when the market rate is 10%, the bond sells at par ($1,000). When rates fall, the market price increases and when rates rise, the market price decreases.

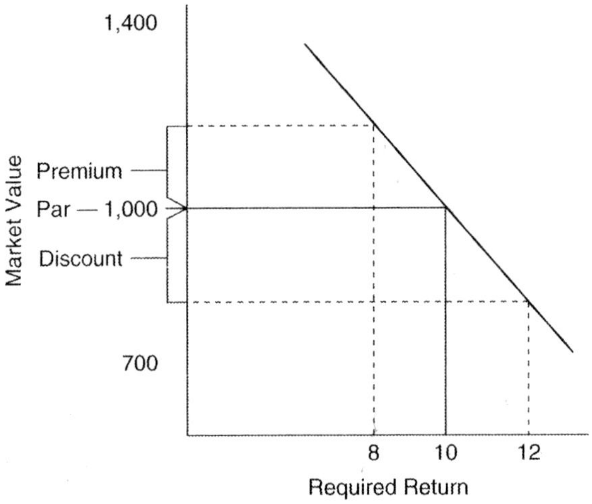

Whenever the required return of a bond is different from its coupon interest rate, the amount of time to maturity affects bond value. Constant required returns allow the bond's value to approach par as the bond moves closer to maturity.

The change in the value of a bond due to changing required returns is affected by time. A change in a bond's required return that has 10 years to maturity will have a greater impact on the bond's value than a change in required return on a bond with only one year left to maturity.

Yield to Maturity (YTM)

LG 6

The yield to maturity (YTM) is the rate that investors earn if they buy a bond at a specific price and hold it until maturity. There are many ways to calculate the YTM of a bond. The trial-and-error approach finds the value of the bond at various rates until the rate causing the calculated bond value to equal its current value is found.

The approximation yield formula is given in the following equation.

$$\text{Approximate Yield} = \frac{I + \dfrac{M - B_0}{n}}{\dfrac{M + B_0}{2}}$$

where:

I = interest payment

B_0 = current market value

M = maturity value

n = periods to maturity

Business calculators and spreadsheet programs provide the most accurate YTM value. Examples follow.

Semiannual Interest and Bond Values

The valuation of a bond paying *semi-annual* interest can be found by performing a simple modification to the equation used to find the value of a bond with annual interest payments.

1. Convert the annual interest, I, to semiannual by dividing by two.

2. Multiply the years to maturity, n, by two.

3. Convert the required stated return from an annual rate, r_d, to a semiannual rate by dividing it by two.

The equation used for finding the value of a bond paying semiannual interest is displayed in the following formula.

$$B_0 = \frac{I}{2} \times \left[\sum_{t=1}^{2n} \frac{1}{\left(1+\frac{r_d}{2}\right)^t} \right] + M \times \left[\frac{1}{\left(1+\frac{r_d}{2}\right)^{2n}} \right]$$

or

$$B_0 = \frac{I}{2} \times \left(PVIFA_{\frac{r_d}{2},2n} \right) + M \times \left(PVIF_{\frac{r_d}{2},2n} \right)$$

Sample Problems and Solutions

Example 1. Valuation of Any Asset

Compute the value of a Picasso oil painting that is projected to sell for \$15 million in 10 years if the cost of capital is 20 percent.

a. Equation Solution

To find the value of any business asset, compute the present value of the cash flows that the asset will generate. In this case, the painting will only generate cash flow when it is sold at the end of 10 years.

$$P_{oil} = \$15,000,000/(1.20^{10}) = \$2,422,580$$

b. Calculator Solution

$N = 10$

$I = 20$

$FV = 15,000,000$

$PMt = 0$

$CPT\ PV = \$2,422,584$

Example 2.

You are appraising a commercial office building that is fully leased and expected to stay that way. The net revenue off the building is \$144,000 per year. The expected life of the building is 30 years. What is the value of the building if the required rate of return is 18%?

a. Equation Solution

Again, all that you need to do is compute the present value of the cash flows. The net cash flows will be \$144,000 per year for 30 years.

$$P_{building} = \$144{,}000 \bullet_{t=1} [1/(1+r_d)^t] = \$144{,}000(5.517) = \$794{,}448$$

b. Calculator Solution

$N = 30$

$I = 18$

$PMT = 144{,}000$

$FV = 0$

$CPT\ PV = \$794{,}420$

Example 3.

How much would you pay for a business that is expected to lose \$10,000 for each of the next five years and then earn \$15,000 each year from then on? Assume a required rate of return of 20 percent.

a. Calculator Solution

PV of Neg CF	PV of perpetuity	PV of Perpetuity at time 0
$N = 5$	15,000/0.2	$N = 5$
$I = 20$	$= 75{,}000_{\text{at time 5}}$	$I = 20$
$PMT = 10{,}000$		$FV = 75{,}000$
$FV = 0$		$PMT = 0$
$CPT\ PV_{\text{neg cash flow}} = -\$29{,}906$		$CPT\ PV_{\text{pos cash flow}} = \$30{,}141$

$PV_{\text{all cash flows}} = -\$29{,}906 + \$30{,}141 = \$235$

Note 1: The present value of the \$15,000 that goes on forever is computed by dividing the cash flow by the discount rate. This is the formula for a perpetuity.

Note 2: Notice that the discount rate used to find the present value of these assets is high. The discount rate reflects the risk of the investment, and because paintings, real estate, and business ventures tend to be risky, it is appropriate to use a high rate when computing value.

Example 4. Bond Valuation

Compute the value of a bond that matures in three years and pays interest annually. The coupon rate is 10 percent and the face amount is $1,000. Assume a discount rate of 12 percent and a discount rate of 8 percent.

a. Calculator Solution

For 12 percent discount rate

N = 3

I = 12

PMT = 100

FV = 1,000

CPT PV = $951.96

For 8 percent discount rate

N = 3

I = 8

PMT = 100

FV = 1,000

CPT PV = $1,051.54

b. Excel Solution

	A	B	C
1	**Bond Value, Annual Interest, Required Return Not Equal to Coupon Interest Rate**		
2	Annual interest payment	$100	$100
3	Coupon interest rate	10%	10%
4	Required return	12%	8%
5	Number of years to maturity	3	3
6	Par value	$ 1,000	$1,000
7	Bond value	$951.96	$1,051.54
8	Entry in Cell B7 is = – *PV*(B4,B5,B2,B6,0)		
9	Entry in Cell C7 is = –*PV*CB4,C5,C2,C6,0)		

Example 5.

Compute the value in Example 4 above, if the bond matures in 20 years instead of in three years.

a. Calculator Solution

N = 20

I = 12

PMT = 100

FV = 1,000

CPT PV = $850.61

Note: When the term to maturity increases, the effect of the increased discount rate is greater. In Example 4, the bond sells for $952.20, but in Example 5, it sells for $850.61.

Example 6.

The following table gives the prices of a bond for various maturities and required returns. The bond pays interest annually and has a 10 percent coupon and a face value of $1,000. Explain what the table implies about interest rate risk.

	Discount Rate		
Term	**9%**	**10%**	**11%**
1	1,008.70	1,000	991.00
10	1,063.00	1,000	940.00
20	1,090.00	1,000	920.30

Solution

As the discount rate increases, the value of the bond falls. This is interest rate risk. If market rates increase after you buy a bond, the value of the bond will decrease. If you sell it, you will suffer a loss. The magnitude of the effect on value of a change in interest rates increases as the term to maturity increases. Notice that a 1 percent increase in the discount rate results in about a $9 decrease in value for the bond that matures in 1 year, but an $80 decrease in value for the bond that matures in 20 years.

Example 7.

A zero coupon bond pays no interest. How much would you pay for a zero coupon bond that pays $1,000 at maturity in 5 years, if the discount rate is 12 percent?

a. Calculator Solution

$N = 5$

$I = 12$

$FV = 1{,}000$

$CPT\ PV = \$567.43$

Example 8.

Compute the value of a bond that matures in 3 years and pays interest *semi-annually*. The coupon rate is 10 percent and the face amount is $1,000. Assume a discount rate of 12 percent.

a. Calculator Solution

$N = 6$

$I = 6$

$PMT = 50$

$FV = 1{,}000$

$CPT\ PV = \$950.83$

Note: Now the number of periods has increased to 6 because there are six payments being made over the three years. The discount has been divided in half as well because the six-month rate will be one half of the annual rate.

Example 9. Yield to Call

Citicorp bonds, which mature in seven years, have a coupon rate of 9 percent, a par value of $1,000, and sell for $904.47. What is the yield to maturity?

a. Equation Solution

To solve for the yield to maturity by hand, use the approximation formula. (The calculator solution is also 11.03 percent.)

$$\text{Approximate Yield} = \frac{I + \dfrac{M - B_0}{n}}{\dfrac{M + B_0}{2}}$$

$$\text{YTM} = \frac{90 + \dfrac{(1000 - 904.47)}{7}}{\dfrac{1000 + 904.47}{2}} = 0.1103$$

b. Calculator Solution

$N = 7$

$PV = -904.47$

$PMT = 90$

$FV = 1{,}000$

$CPT\ I = 11.03$

c. Excel Solution

	A	B
1	**YIELD TO MATURITY, ANNUAL INTEREST**	
2	Annual interest payment	$90
3	Coupon interest rate	9%
4	Current bond price	– 904
5	Number of years to maturity	7
6	Par value	$1,000
7	Yield to Maturity	11.03%
8	Entry in Cell B7 is = RATE(B6,B3,B5,B7,0)	

Example 10. Real and Nominal Rates of Interest

You currently have $1,000 that you could spend today to purchase concert tickets costing $100 each, or you could invest the $1,000 in a risk-free U.S. Treasury security earning 4 percent nominal rate of interest. The consensus forecast of leading economists is a 2 percent rate of inflation over the coming year.

a. How many concert tickets could you purchase today?

b. How much money would you have at the end of one year if you do not purchase the concert tickets but invest the $1,000 in the U.S. Treasury?

c. How much would you expect the concert tickets to cost in one year?

d. How many concert tickets could you purchase in one year (fractions are okay)? In percentage terms, how many more or fewer concert tickets can you buy at the end of year 1?

e. What is your real rate of return over the year? How is the real rate of return related to the percentage change in your buying power found in part d?

Solution:

a. You could buy 10 concerts tickets today. $1,000/$100 = 10

b. You would have $1,040 in one year if you invested the $1,000. $1,000 × (1.04) = $1,040

c. Concert ticket would cost $102 in one year because of inflation. $100 × (1.02) = $102

d. You would be able to purchase 10.2 concert tickets in one year. $1,040/$102 = 10.2. In percentage terms, you would be able to purchase 2 percent more tickets at the end of year one. (10.2/10) – 1 = 0.02 = 2%.

e. The real rate of return is the risk-free rate minus the inflation rate, which is 2 percent (4% – 2%). The change in the number of tickets that can be purchased is determined by the real rate of return since the portion of the nominal return for expected inflation (2 percent) is available just to maintain the ability to purchase the same number of tickets.

■ Quick Drill Self Test

These problems are quick and easy practice to see how well you have understood the concepts in this chapter. The answers follow.

1. A building has an expected life of 30 years. Its net cash flows are projected to be $120,500 per year. If a discount rate of 17 percent is appropriate, what is the value of the building?

2. If the discount rate were to increase to 23 percent because of a perception of increased risk in the real estate market, what would be the revised value of the building?

3. A bond's par value is $1,000. It has five years until maturity. Its coupon rate is 7 percent. What is the value of the bond if the market rate is 10 percent, assuming annual compounding?

4. Given the bond in Problem 3, what would the current value be if the market rate were 5 percent?

5. Again, looking at the bond in Problem 3, what would the value be if it had 30 years to maturity instead of five years, assuming all other features are as given in Problem 3?

6. What is the price of a bond with features as are given in Problem 3 except that it matures in one year rather than in five years?

7. Review your solutions to Problems 5 and 6. Which bond is subject to the greatest interest rate risk?

8. What would the value of the bond discussed in Problem 3 be if interest payments made were semiannual rather than annually?

9. What would be the value of the bond discussed in Problem 4 if interest payments were made semiannually rather than annually?

10. What will be the price of the bond discussed in Problem 3 when there are four years before it matures?

11. What is the percentage change in price of the bond discussed in Problem 10 between when there are five years before it matures and when there are 4 years before it matures?

12. What is the current yield of the bond discussed in Problem 3?

13. Is the sum of the current yield and the change in price found in Problem 11 the same as the market rate?

14. A bond has a current market price of $1,125. It has a annual coupon rate of 6 percent, a par value of $1,000, and matures in 10 years. What is its yield to maturity?

15. A bond has a current market price of $825. It has an annual coupon rate of 6 percent, a par value of $1,000, and matures in 10 years. What is its yield to maturity?

16. What would be the yield to maturity for the bond discussed in Problem 14 if the payments were made semiannually rather than annually? If you use a calculator, remember to multiply your answer by 2.

17. What would be the yield to maturity for the bond discussed in Problem 15 if the payments were made semiannually rather than annually?

Answers to Quick Drill Self Test

1. $702,441.45
2. $522,860.82
3. $886.28
4. $1,086.59
5. $717.19
6. $972.73
7. Problem 5
8. $884.17
9. $1,087.52
10. $904.90
11. 2.10 percent
12. 7.90 percent
13. Yes
14. 4.43 percent
15. 8.69 percent
16. 4.44 percent
17. 8.65 percent

■ Study Tips

1. The principle behind valuation is simple. Any financial asset may be valued by finding the present value of all cash flows. The trick is (a) identifying the cash flows, and (b) determining how to compute their present value. This later issue requires a thorough understanding of time-value-of-money mathematics. If you are struggling with this chapter, go back and reread the TVM chapter, paying particular attention to the sections covering present value.

2. Financial calculators are especially useful for bond calculations.

3. You can check your work on bond valuation problems by applying the discount/premium method. You learned that as interest rates fall, the price of the bond increases, and vice versa. When you value a bond, compare the discount rate to the coupon rate and decide whether the bond should be selling for more or less than $1,000. Compare your answer and see if it is reasonable. For example, say you are to find the value of a bond with a 10 percent coupon, five years to maturity, and an 8 percent discount rate. Because the discount rate is below the coupon rate, rates have fallen. When rates fall, the bond price rises. Therefore you know that the value of the bond must be more than $1,000. In fact, the value of this bond is actually $1,079.85.

■ Student Notes

____

____

____

____

■ Sample Exam—Chapter 6

True/False

T F 1. The value of any asset depends on the accounting profits it is expected to provide over the ownership period.

T F 2. If the value of a bond, B_0, is less than its par value, the bond is selling at a discount.

T F 3. The yield to maturity on a bond with a current price equal to its par value will always equal the coupon interest rate.

T F 4. A bond covenant contains a description of the call feature and the sinking fund for a bond.

T F 5. The bond indenture reports on the risk characteristics of the bond.

T F 6. Bond rating agencies provide investors a rating that reflects the default risk of a bond issue.

T F 7. Any action that increases the risk of an asset will also increase that asset's required return.

T F 9. Bonds with short maturities will have less interest rate risk than bonds with longer maturities and equal features.

T F 10. As market interest rates increase, bond prices increase.

T F 11. The discount rate of a bond is synonymous with the bond's required return.

T F 12. The trial-and-error approach to bond valuation involves finding the value of the bond at various yield-to-maturity rates until the rate that causes the calculated bond value to equal its current value is found.

T F 13. The longer the period until a bond matures, the less a change in market interest rates will affect its market value.

T F 14. Longer-term bonds have lower interest rate risk than short-term bonds.

T F 15. The value of a bond is the present value of its coupon payments until maturity plus the present value of its par value at maturity.

T F 16. The nominal rate of interest is equal to the real rate of interest plus the inflation premium plus the risk premium.

T F 17. The term structure of interest rate is the relationship between the risk and the return for bonds.

Multiple Choice

1. _________ is the process that links risk and return in order to determine the worth of an asset.
   a. Depreciation
   b. Valuation
   c. Evaluation
   d. Discounting

2. The value of an asset is the _________ of all future cash flows it is expected to provide over a relevant time period.
   a. required return
   b. future value
   c. present value
   d. sum

3. Risk is generally incorporated into the _________ in the present value model.
   a. discount rate
   b. cash flows
   c. timing
   d. total value

4. A bond will sell at _________ if the required return is greater than the coupon rate.
   a. par
   b. a premium
   c. liquidation value
   d. a discount

5. The present value of a bond's _________ determines the value of the bond.
   a. dividend
   b. coupon payment
   c. maturity value
   d. coupon payment and maturity value

6. _________ is/are long-term debt instruments used by business and government.
   a. Preferred stock
   b. Retained earnings
   c. Bonds
   d. Common stock

7. Interest rate risk and the time to maturity have a relationship that is best characterized as
   a. inverse.
   b. unrelated.
   c. varying.
   d. direct.

8. What is the approximate yield to maturity for a $1,000 par value bond selling for $1,100 that matures in 5 years and pays a 10 percent coupon?
   a. 12.5 percent
   b. 7.6 percent
   c. 10 percent
   d. 13.6 percent

9. As market rates increase the value of a bond will ________, all other things equal.
   a. increase
   b. decrease
   c. remain unchanged
   d. be undeterminable

10. The longer the term a bond has before it matures, ________ will be the effect on its value due to a 1 percent change in market interest rates.
    a. the greater
    b. the lower
    c. Either a or b is correct, depending on the direction of the interest rate change.
    d. Neither a nor b becauses bond maturity has no impact on interest rate sensitivity.

11. What is the value of a bond with 10 years to maturity, a 7 percent coupon rate, annual interest payments, $1,000 par, and a 7 percent market rate?
    a. $900
    b. $1,000
    c. $1,100
    d. $1,150

12. The yield to maturity on a bond with a price equal to its par value will
    a. always be equal to the coupon rate.
    b. will be more than the coupon rate.
    c. will be lower than the coupon rate.
    d. will depend upon the required return.

13. What is the yield to maturity for the following bond (to the nearest full percentage)?

Current Price	Coupon Rate	Par Value	Frequency of Interest Payment	Years to Maturity
$954	12%	$1,000	Annually	7

    a. 7 percent
    c. 12 percent
    b. 9 percent
    d. 13 percent

14. For an investor who plans to purchase a bond that matures in one year, the primary concern should be
    a. yield to maturity.
    c. interest rate risk.
    b. coupon rate risk.
    d. exchange rate risk.

15. A bond pays no interest. It has a $1,000 par and matures in five years. What is its value if market rates are 10 percent?
    a. $0
    b. $500
    c. $1,000
    d. $620.92

16. The key inputs to the valuation process include cash flows (returns), the required return (risk), and
    a. discount rates.
    c. timing.
    b. present value.
    d. expected return.

17. Corporate bonds usually have which of the following?
    a. face value of $1,000
    b. an initial maturity of 10 to 30 years
    c. semiannual interest payments
    d. all of the above

18. A firm has issued a bond that has a coupon of 11 percent, semiannual interest payments, a required return of 14 percent, and 10 years left until maturity. What will the bond sell for today?
    a. $1,000
    c. $843
    b. $841
    d. $895

19. A bond has a current market price of $1,100. It is said to be selling at a
    a. premium.
    b. discount.
    c. bonus.
    d. reward.

20. Which of the following is not likely to be included in bonds' covenants?
    a. restriction on dividends
    c. minimum levels of liquidity
    b. limits on additional debt
    d. call provision

21. In early 2007, three-month Treasury bills were offering a return of approximately:
    a. 3 percent.
    b. 5 percent.
    c. 7 percent.
    d. 9 percent.

22. All of the following are measures of total return, except
    a. current yield.
    b. yield to call.
    c. yield to maturity.
    d. Trick question, all are measures of total return.

23. Which of the following bonds has the highest cost per unit?
    a. Bond A, a $2,000 par value bond priced at $72.045
    b. Bond B, a $1,000 par value bond priced at $100.000
    c. Bond C, a $900 par value bond priced at $144.090
    d. You would pay the same amount for Bond A and Bond C.

24. A firm has issued a bond that has a coupon of 9 percent, semiannual interest payments, a current price of $938.00 and 10 years left until maturity. What is the bond's yield to maturity?
    a. 5 percent
    b. 10 percent
    c. 9 percent
    d. 7 percent

25. If a AAA rated corporate bond has a required return of 6.5 percent, the risk-free rate is 5 percent, and the inflation premium is 3 percent, then what is the risk premium for the bond?
    a. 6.5 percent
    b. 5 percent
    c. 3.5 percent
    d. 1.5 percent

Essay

1. What is interest rate risk? When is it greatest?

2. Under what conditions will an investor actually receive a return on a bond equal to the yield to maturity?

3. Why do required returns on bonds change over time?

Chapter 6 Answer Sheet

True/False

1. F
2. T
3. T
4. F
5. F
6. T
7. T
8. F
9. T
10. F
11. T
12. T
13. F
14. F
15. T
16. T
17. F

Multiple Choice

1. B
2. C
3. A
4. D
5. D
6. C
7. D — $$\text{YTM} = \frac{100 + \dfrac{1{,}000 - 1{,}100}{5}}{\dfrac{1{,}000 + 1{,}100}{2}} = 7.62\%$$
8. B
9. B
10. A
11. B
12. A
13. D — $N = 7$, $FV = 1000$, $PMT = 120$, $PV = -954$, $CPT\ I = 13.04$
14. A
15. D
16. C
17. D
18. B — $N = 20$, $I = 7$, $PMT = 55$, $FV = 0$, $CPT\ PV = 843$
19. A
20. D
21. B
22. A
23. A
24. B — PV = –938, N = 20, FV = 1000, PMT = 45, CPT I = 5% (semi) 5% × 2 = 10% (YTM)
25. D

Essay

1. Interest rate risk is the volatility in the price of a bond due to changing interest rates. It is greatest when the time to maturity is longest.

2. An investor will earn the yield to maturity only if the bond is held until it matures. The yield to maturity includes both interest income and changes in the value of the bond over time. The current yield is computed as the interest payment divided by the current market price. The yield to maturity will include this return plus any return from price changes in the bond over time. If the bond is not held until it matures, these price changes will not be fully received.

3. Required returns of bonds change over time because of changes in the economy causing all nominal rates to increase or decrease, such as a change in inflation expectations, and because the underlying risk of a bond itself may change causing a change in the risk premium.

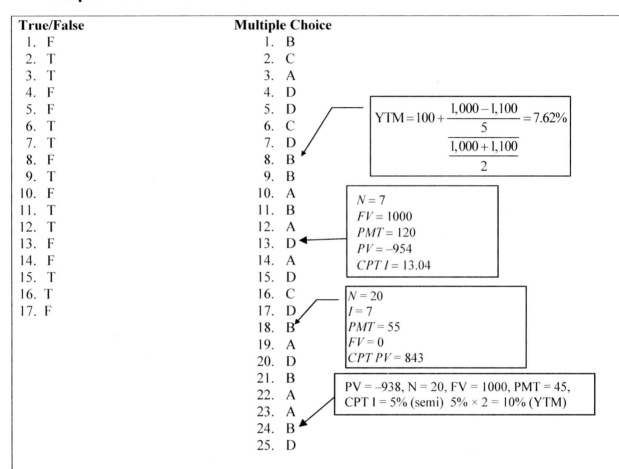

Chapter 7
Stock Valuation

■ Chapter Summary

As we discussed in Chapter 6, all assets are valued as the present value of all future cash flows. The value of a bond is the present value of its interest and principal payments and the value of a building is the present value of its net cash flows. The value of stock is also found as the present value of its cash flows. The problem is that estimating these cash flows can be very difficult. We will investigate various methods for estimating cash flows and for computing their value in this chapter.

LG 1 *Differentiate between debt and equity capital.* Capital refers to a firm's long-term sources of funds. This capital can be either borrowed or equity. Debt includes all borrowing incurred by the firm. Equity consists of funds provided by the firm's owners (investors or stockholders). A firm can obtain equity either internally, by retained earnings, or externally, by selling common or preferred stock.

LG 2 *Discuss the features of both common and preferred stock.* Common stock is the type of stock you usually think of when discussing stock. It gives the holder the right to residual earnings and the right to vote in key company issues. Preferred stockholders neither vote nor share in company profits. They only receive a fixed dividend payment.

LG 3 *Describe the process of issuing common stock, including venture capital, going public, and the investment banker.* Most firms do not raise equity or issue stock very often, so they must enlist the services of various professionals, such as venture capitalists and investment bankers, to help with the process.

LG 4 *Understand the concept of market efficiency and basic common stock valuation using zero-growth, constant-growth, and variable-growth models.* The idea behind market efficiency is that all securities are priced correctly all of the time. If a stock was overpriced, investors would sell it. This would reduce its price and increase its return. When the return has risen to its correct level, the selling will stop. Stocks are valued based on the present value of their dividends. For firms that pay a constant dividend, the perpetuity formula is used. This chapter introduces other simplifying methods for valuing stock.

LG 5 *Discuss the free cash flow valuation model and the book value, liquidation value, and price/earnings (P/E) multiples approaches.* The free cash flow model determines the value of an entire company as the present value of its expected free cash flows discounted at the firm's weighted average cost of capital, which is its expected average future cost of funds over the long run. The book and liquidation value methods to stock valuation ignore the value of future cash flows. The P/E method involves multiplying the firm's expected earnings per share by the industry P/E.

LG 6 *Explain the relationships among financial decisions, return, risk, and the firm's value.* Managerial decisions that decrease cash flows or increase risk will result in low stock prices. Similarly, if management can increase cash flows without increasing risk, stock prices can be increased. This is the goal of financial managers.

■ Chapter Notes

Valuation of stock is complicated by the fact that the future cash flows are not known with any certainty. This chapter discusses various ways analysts value stock despite this problem.

Debt versus Equity Capital

LG 1

In the last chapter we discussed the characteristics of debt. This chapter focuses on equity. Investors who buy equity are called stockholders, and they receive certain rights. These rights include:

1. A voice in management
2. Claims on income and assets

Stock can be distinguished from debt by two additional features:

1. The security does not mature.
2. Dividend payments to stockholders are not tax deductible.

The true owners of the firm are the stockholders. They are sometimes referred to as the residual owners because they receive whatever is left after all other claims on the business have been satisfied. Stockholders can share in the good fortunes of the firm if it does well. However, stockholders are not guaranteed anything except they cannot lose more than they have invested.

Features of Common and Preferred Stock

LG 2

Terminology

Par value: This is a useless term that refers to the value assigned to one share of stock for accounting purposes—usually $1.00.

Preemptive rights: This allows common stockholders to maintain their proportionate ownership in the corporation when new shares are issued by giving them the right to buy more shares in any future issue.

Authorized, outstanding, and issued shares: The firm's charter will specify the maximum number of shares it may issue. More shares can only be issued if shareholders agree by a vote. The number of shares outstanding is the number that is held by the public. Issued shares are the shares of common stock that have been put into circulation: They represent the sum of outstanding shares and treasury stock.

Voting rights: Stockholders have the right to vote on key issues affecting the firm.

Dividends: Stockholders have the right to receive dividends at the discretion of the board of directors, but dividends are not contractual. Dividends are usually paid quarterly.

American Depository Shares: Securities backed by stocks of foreign companies that are held in bank vaults that facilitate U.S. investor purchase of non-U.S. companies and trade them in U.S. markets.

Issuing stock outside of their home markets can benefit corporations by broadening the investor base and also allowing them to become better integrated into the local business scene. A local stock listing both increases local press coverage and serves as effective corporate advertising. Locally traded stock can also be used to make corporate acquisitions.

Preferred Stock

Preferred stock is as much like debt as it is like common stock. Preferred stockholders receive a fixed dividend payment that does not change, even if the firm does extremely well. They are not allowed to vote unless the firm has defaulted on its dividends payments. It is called preferred stock because preferred stock dividends are paid before common stock dividends.

Preferred stock is most often used by public utilities that want to increase their financial leverage without increasing risk, by acquiring firms in merger transactions, or by firms that are experiencing losses and need additional financing. Preferred stock is frequently combined with other special features, such as conversions or warrants, to enhance its attractiveness to investors.

Basic Rights of Preferred Stockholders

Distribution of earnings: Preferred stockholders are given priority over common stockholders with respect to the distribution of earnings. If the stated preferred stock dividend is passed (not paid) by the board of directors, the payment of dividends to common stockholders is prohibited.

Distribution of assets: Preferred stockholders are usually given preference over common stockholders in the liquidation of assets as a result of a firm's bankruptcy, although they must wait until all creditors have been satisfied.

Voting rights: Because preferred stockholders are not exposed to the same amount of risk as common stockholders (distribution of earnings, distribution of assets, fixed dividend payment), they are not normally given voting rights.

Features of Preferred Stock

Restrictive covenants: Restrictive covenants are usually placed on preferred stock to ensure the continued existence of the firm and the regular payment of the stated dividend. Violation of a preferred stock covenant usually permits preferred stockholders either to obtain representation on the firm's board of directors or to force the retirement of their stock at or above its par, or stated value.

Cumulative: For preferred stock that is *cumulative*, all dividends in arrears must be paid along with the current dividend before payment of dividends to common stockholders. Preferred stock that is *noncumulative* does not accumulate dividends.

Participation: Most issues of preferred stock are *nonparticipating*, which means that preferred stockholders receive only the specified dividend payments. Occasionally, *participating* preferred stock is issued, which allows preferred stockholders to participate with common stockholders in the receipt of dividends beyond a specified amount.

Call feature: Preferred stock is usually callable, which means that the issuer can retire outstanding stock within a certain period of time at a specified price. The call price is usually set above the initial issuance price, but it may decrease according to a predetermined schedule as time passes.

Conversion feature: Preferred stock often contains a conversion feature that allows holders to convert each share into a stated number of shares of common stock.

Preferred Stock Advantages

Ability to increase financial leverage: Because preferred dividends are fixed payments to its holders, the use of preferred stock will help to increase the firm's financial leverage, which will magnify the effects of increased earnings on common stockholders' returns.

Flexibility: Although preferred stock is similar to debt, it differs from debt in that the issuer can pass a dividend payment without suffering the consequences that result when an interest payment is missed on a bond.

Use in corporate restructuring: Preferred stock is often exchanged for shares of common stock in the restructuring process of a corporation. The preferred stock allows for the payment of a fixed dividend and for the remaining funds to be reinvested, perpetuating the growth of the new enterprise.

Preferred Stock Disadvantages

Seniority of the preferred stockholder's claim: Preferred stockholders are given priority over common stockholders but fall behind creditors in a legally bankrupt firm, which jeopardize the returns of common stockholders.

Cost: Unlike the payment to debt holders, the payment to preferred stockholders is not guaranteed, and the holders of preferred stock must be compensated for this risk with higher interest rates. Also, preferred stock dividends are not tax deductible, unlike the interest payments on debt.

Difficult to sell: A large number of investors find preferred stock unattractive, and most often firms must include special features such as conversion or warrants to enhance its marketability.

Process of Issuing Common Stock

LG 3

Most firms begin with an initial investment by its founders in the form of common stock. As the firm develops, angel capital and/or venture capital firms may supply additional funds. These firms look for opportunities to get involved with companies that are in their early stage of product development. Venture capitalists typically are formal business entities that maintain strong oversight over the firm, while angel capitalists tend to be investors who do not operate as a business but are often groups of wealthy individuals willing to invest. Although many startup firms fail, when one is successful the returns can be very high.

There are four ways in which institutional venture capitalists are most commonly organized.

- *Small business investment companies* (*SBICs*) are corporations chartered by the federal government.
- *Financial VC funds* are subsidiaries of financial institutions, particularly banks.
- *Corporate VC funds* are firms, sometimes subsidiaries, established by nonfinancial firms.
- *VC limited partnerships* are limited partnerships organized by professional *VC* firms, who serve as general partner.

The term "going public" refers to selling company common stock to the public for the first time. The firm may make a (1) public offering, (2) a rights offering, or (3) a private placement.

The first time stock is offered to the public is called an initial public offering (IPO). This is a primary market transaction because the issuing firm is receiving the proceeds of the sale. An investment banking firm will usually be involved in an IPO. The main activity of the investment banker is underwriting. Underwriting involves purchasing the stock from the issuing firm and reselling it to the public, usually through its brokerage network. If the issue is large the investment banking firm may form a syndicate to spread the risk. A syndicate is a group of investment banking firms that join together temporarily to market a security to the public.

Common Stock Valuation

LG 4

To determine the value of a common stock, it is important to keep in mind the concept of market efficiency. The prices of common stocks are constantly adjusting to the differences in investors' required returns, *k*, and the investors' expected return, *K*. The expected return is the return that is expected to be earned each period on a given asset over an infinite time horizon. If the expected return is less than the required return, investors will sell the asset because the return is not sufficient for the amount of risk that is involved. This will drive the price down to the point where the required return equals the expected return. The opposite is true for an asset that has an expected return greater than its required return. Investors will buy the asset, driving its price up to the point where the expected return equals the required return.

According to the efficient market hypothesis:

a. Securities prices are in equilibrium (fairly priced with expected returns equal to required returns).
b. Securities prices fully reflect all public information available and will react quickly to new information.
c. Investors should therefore not waste time searching for mispriced (over- or undervalued) securities.

The efficient market hypothesis is generally accepted as being reasonable for securities traded on major exchanges; this is supported by research on the subject. There is an increasing challenge to the efficient market hypothesis being offered by the study of finance behavior. The challenge comes primarily from the fact that tests of the efficient market hypothesis assumes investors are completely rational. A going body of research disputes this rationality assumption and shows that investors are driven by the irrational behaviors of greed, fear, and other emotions.

The Basic Stock Valuation Equation

The value of common stock is equal to the present value of all future dividends it is expected to provide over an infinite time horizon. From a valuation viewpoint, capital gains are not considered; only dividends are taken into account. The basic stock valuation equation is as follows:

$$P_0 = \frac{D_1}{(1+r_s)^1} + \frac{D_2}{(1+r_s)^2} + \cdots + \frac{D_\infty}{(1+r_s)^\infty}$$

where:

P_0 = value of common stock

D_t = per-share dividend expected at the end of year t

r_s = required return on common stock

The equation can be simplified by redefining each year's anticipated dividend growth. For stocks that have dividends with zero growth, the following zero-growth model can be used.

$$P_0 = \frac{D_1}{r_s}$$

The constant growth model, also referred to as the Gordon model, assumes that dividends will grow at a constant rate, g, which is less than the required return. To find the growth rate of a stock's dividend, simply apply the technique described in Chapter 5 for finding growth rates. In its simplified state, the constant growth model can be written as follows:

$$P_0 = \frac{D_1}{r_s - g}$$

where: g = constant growth rate

The variable growth model assumes a shift in the growth rate, g, after n years. To use the model, one must find the dividends at the end of each year during the initial growth period. Next, find the present value of dividends occurring within the initial growth period and sum these values to obtain the net present value of the dividends during the initial growth period. Next, find the value of the stock at the end of the initial growth period by multiplying the dividend of the last year of initial growth by the new growth rate. Plug the new dividend into the constant growth model to find the stock's value. Finally, find the present value of the stock in the new growth period and add it to the present value of its dividends in the initial growth period to determine the stock's value.

Example

Determine the value of a stock which had a recent dividend payment of $1.00 in 2013. The dividend is expected to grow at a 10 percent rate for the next two years. After 2 years, the growth rate is expected to drop to 5 percent. The firm has a 12 percent required return.

t	End of Year	Dividend in Year 0	$(1 + r)^t$	Dividend at the End of Year t	$1/(1 + r)^t$	Present Value of Dividends
1	2014	$1.00	1.100	$1.10	0.893	0.98
2	2015	$1.00	1.210	$1.21	0.797	0.96
				Sum of PV of dividends in 2011 and 2012 = $1.94		

$D_{2016} = D_{2015} \times 1.05 = \$1.21 \times 1.05 = \$1.27$

$P_{2015} = [\$1.27/(0.12 - 0.05)] = \$18.14$ (Present value of dividends from 2016 to infinity at year 2015)

$\$18.14 \div (1.12)^2 = \$18.14 \times 0.797 = \$14.46$ (Present value today of dividends from 2016 to infinity)

$P_{2013} = \$1.94 + \$14.46 = \$16.40$

Free Cash Flow Valuation and Other Approaches to Common Stock Valuation

LG 5

The free cash flow model determines the value of an entire company as the present value of its expected free cash flows discounted at the firm's weighted average cost of capital, which is its expected average future cost of funds over the long run.

$$V_C = \frac{FCF_1}{(1+r_a)^1} + \frac{FCF_2}{(1+r_a)^2} + \cdots + \frac{FCF_\infty}{(1+r_a)^\infty}$$

where

V_C = value of the entire company

FCF_t = free cash flow *expected* at the end of year t

r_a = the firm's weighted average cost of capital

Because the value of the entire company, V_C, is the market value of the entire enterprise (that is, of all assets), to find common stock value, V_S, we must subtract the market value of all of the firm's debt, V_D, and the market value of preferred stock, V_P, from V_C.

$$V_S = V_C - V_D - V_P$$

The book value per share is the amount per share of common stock to be received if all of the firm's assets are sold for their exact book value and if the proceeds remaining, after all liabilities are paid, are distributed among the common stockholders. This method is not often used because it ignores the firm's earning potential.

The liquidation value per share is the actual amount per share of common stock to be received if all of the firm's assets are sold, all liabilities are paid, and any remaining money is divided among the shareholders. The liquidation value per share is more accurate than the book value approach but it still fails to recognize future earnings.

The price/earnings multiple approach is calculated by multiplying the firm's expected earnings per share (EPS) by the average price/earnings (P/E) ratio for the industry. The price/earnings multiple approach should be used before the book value or liquidation value approach because it takes into account the expected earnings of the firm.

Decision Making and Common Stock Value

LG 6

Stock valuation is based on expected returns and risk. Changes in these variables can have a substantial impact on the valuation process.

Assuming that economic conditions remain constant, an increase in stockholders' dividend expectations should raise the value of the firm. If a financial manager can increase the level of expected returns without increasing risk, the value of the firm will increase.

Any action taken by the financial manager that increases risk will also increase the required return. An increase in the required return will lower the share value, P_0, of a firm. Any decision made by the financial manager that increases risk will reduce the share value of a firm. The same holds true of any reduction in risk. With reduced risk, share value will increase.

Any decision made by the financial manager rarely affects return and risk independently. Keep in mind that the net result of the financial manager's decision should be an increase in shareholder wealth.

■ Sample Problems and Solutions

Example 1. Stock Valuation

Assume the current dividend is \$3.00 per year. The firm has grown at a constant rate of 5 percent for the last 10 years, and this growth rate is expected to continue. If the appropriate discount rate is 14 percent, what is the value of the stock?

Solution

Use the constant growth model, $P = [D_0(1 + g)]/(r_s - g)$, where P = price of the stock, D_0 = the current dividend, r = discount rate, and g = growth rate.

$$\text{price stock} = \frac{\$3.00(1.05)}{0.14 - 0.05} = \$35$$

Example 2.

Assume that *next period's* dividend is expected to be \$3.15. If a growth rate of 5 percent is assumed and the appropriate discount rate is 14 percent, what is the value of the stock?

Solution

The usual error that students make using the constant growth model is getting confused with what goes in the numerator. If next period's dividend is given in the problem, then the formula is

$$P = D_1/(r_s - g)$$

where D_1 = *next period's* dividend.

$$\text{price stock} = \frac{\$3.15}{0.14 - 0.05} = \$35$$

Example 3. Zero Growth Model

What is the price of a share of stock for a firm that pays a \$2.00 dividend? The dividend has never changed and is not expected to in the future. The appropriate discount rate is 12 percent.

Solution

If the dividends are expected to stay the same forever, then the zero growth model can be used. After reviewing the constant growth model, what would you expect the zero growth model to be? If the growth rate is zero, then the constant growth model becomes the formula for a perpetuity, which is what we would expect.

$$\text{price stock} = \$2/0.12 = \$16.67$$

Example 4. Mixing It Up

A more realistic example than the constant growth or zero growth model is what might be called an uneven growth. Suppose that you are evaluating the stock of a relatively new firm. For the first several years you expect high growth, say 20%. Then a more moderate level of growth may be observed. Finally, the dividends may be constant when the firm has matured. The principles used to work out the value of this stock are the same as used in the other examples. That is, find the present value of the cash flows. Begin by drawing a time line, then compute the present value of each cash flow and, finally, add them all together.

A firm is expected to grow at an annual rate of 20% for the next 2 years. The growth is expected to decrease to 8% for the following 2 years and to be 4% thereafter. Assuming the current dividend is \$2.00/yr and that the appropriate discount rate is 15%, what would be the value of the stock?

Solution

First draw a time line that shows what the dividends are going to be each year.

2	2.4	2.88	3.11	3.36	3.49	. . . constant growth at 4%	
0	1	2	3	4	5	6	7

Next compute the present value of the first four dividends.

$$PV = 2.4/1.15 + 2.88/1.15^2 + 3.11/1.15^3 + 3.36/1.15^4 = \$8.23$$

Now find the present value of the constant growth portion of the cash flow.

$$PV = 3.49/(0.15 - 0.04) = \$38.78/(1.15^4) = \$22.17$$

Finally, add the two parts together.

$$\text{Price of stock} = \$8.23 + \$22.17 = \$30.40$$

***Note*:** The most common mistake in these types of problems is to not discount the growth portion of the cash flow for the right number of periods. Recognize that the constant growth model finds the value one period before the dividend in the numerator is received. This is at the end of period 4. That means that the \$38.78 must be discounted back four periods to find its value at time zero.

Exercise 5. Decision Making and Common Stock Value

Ryan International, Inc.'s most recent dividend was \$2.50 per share, its expected annual rate of dividend growth is 7%, and its required return is now 12 percent. Management is considering merging with another firm, which is expected to increase growth to 8 percent and increase the required return to 13 percent. Should Ryan International merge or not? Why?

Solution

The current value of Ryan International before the merger is \$53.50 [(2.50 × 1.07)/(0.12 – 0.07) = 53.50]. If Ryan merges the value of the firm is expected to be \$54 [(2.50 × 1.08)/(0.13 – 0.08) = 54]. Since the value for the firm is expected to increase with the merger, Ryan International should merge with the other firm.

■ Quick Drill Self Test

These problems are quick and easy practice to see how well you have understood the concepts in this chapter. The answers follow.

1. What is the price of a share of preferred stock that has a dividend of \$3 if the cost of preferred (r_p) is 15 percent?

2. What is the value of a share of preferred stock if the dividend is \$3 and the price is \$30?

3. What would you pay for a share of common stock if you expect the next dividend to be \$2 and you expect to sell it in one year for \$20, assuming the cost of equity (r_e) is 10 percent?

4. What would you pay for a share of common stock if you expect the next dividend to be \$2 and you expect to sell it in one year for \$20, assuming the cost of equity (r_e) is 15 percent?

5. What would you pay for a share of stock if the next dividend was \$2, the one after that was \$2.5, and after two years you expected to sell the stock for \$15, assuming a cost of equity of 15 percent?

6. What would you pay for a share of common stock with a \$3 dividend that was not expected to grow in the future, assuming a 15 percent cost of equity?

7. What would you pay for a share of common stock where the last dividend was \$3 and that was expected to grow at 5 percent per year indefinitely, assuming a 15 percent cost of equity?

8. What would you pay for a share of common stock where the last dividend was \$3 and that was expected to grow at 8 percent per year indefinitely, assuming a 15 percent cost of equity?

9. What would you pay for a share of common stock where the last dividend was \$3 and that was expected to grow at 5 percent per year indefinitely, assuming a 25 percent cost of equity?

10. What would you pay for a share of common stock where the last dividend was \$3 and that was expected to grow at –5 percent per year indefinitely, assuming a 25 percent cost of equity?

11. You do not expect there to be a dividend paid next year. After that, however, you expect a \$2 dividend that will grow at a constant 6 percent thereafter. How much would you pay for the stock assuming a cost of equity of 10 percent?

12. What will be the price of the stock discussed in Problem 11 one year from now?

13. What is the percentage increase in price between now and one year from now using the results found in Problems 11 and 12?

14. A stock currently sells for \$25 and pays a \$3 dividend. What is the current yield?

15. Suppose the current yield on a stock was 3 percent and the cost of equity was 15 percent. What would you expect the appreciation in the stock's price to be?

16. The beta of a stock is 1.5. The required return on the market is 12 percent and the risk free rate is 5 percent. What is the cost of equity?

17. What is the price of a share of stock if the beta is 2, its next dividend is projected to be \$3 and its growth rate is expected to be a constant 7 percent, assuming the market return is 15 percent and the risk free rate is 6 percent?

18. The PE ratio for a firm is 20. Expected earnings per share are \$1.5. What is the current price of the firm's stock?

19. You wish to estimate the value of a company that has not yet issued stock. The PE ratios of similar firms is 16. You expect earnings to be \$200,000. What is the total value of the firm?

20. Given the data in Problem 19, what would be the price per share if you issue 100,000 shares?

Answers to Quick Drill Self Test

1. $20
2. 10 percent
3. $20.00
4. $19.13
5. $14.97
6. $20.00
7. $31.50
8. $46.29
9. $15.75
10. $9.50
11. $45.45
12. $50.00
13. 10 percent
14. 12 percent
15. 12 percent
16. 15.5 percent
17. 17.65
18. $30.00
19. $3,200,000
20. $32

■ Study Tips

1. The Gordon growth model is really a fairly simple equation to use; however, students tend to make errors using it by being sloppy about counting time periods. The Gordon growth model finds the value of stock with constant growth one period before the time when the dividend in the numerator occurs. For example, if the dividend used in the numerator is received at the end of period 5, the Gordon growth model provides the value of the stock at the end of period 4. Review the examples in this study guide and make sure that you understand this point.

2. The mathematics of solving for the price of stock with odd growth at the beginning followed by constant growth is about as difficult as you will encounter in this course. Spend the time it takes to fully understand how to do these problems. This is where your work on time value of money begins to pay off.

3. If your calculator has the ability to handle multiple cash flows you should learn how to apply this to solving these problems.

■ Student Notes

■ Sample Exam—Chapter 7

True/False

T F 1. It is easier to value stock than it is to find the value of bonds.

T F 2. The value of a share of stock depends on the cash flows investors can expect to receive by owning that stock.

T F 3. The dividend valuation model assumes that dividends continue being paid forever.

T F 4. When valuing common stock, capital gains (as opposed to dividends) determine a majority of the stock's current value.

T F 5. When using the constant growth model for the valuation of common stock, the constant growth rate, g, has to be greater than the required return, r_s, in order for the model to be useful.

T F 6. The book value of an asset is simply the amount paid for the asset minus its accumulated depreciation.

T F 7. Any action that increases the risk of an asset will also increase that asset's required return.

T F 8. Under the efficient market hypothesis, when the expected return of a common stock is less than its required return, investors will buy the stock and drive the price upward.

T F 9. The constant growth model requires that required return exceed long term expected growth.

T F 10. The most accurate approach to common stock valuation is the book value per share method.

T F 11. The numerator in the constant growth model is NEXT period's expected dividend.

T F 12. The price earnings approach to stock valuation relies on the market accurately setting PEs that reflect firm value.

T F 13. The variable growth model used in the valuation of common stock can be used to find the value of preferred stock.

T F 14. The price/earnings multiple approach to common stock valuation takes a firm's expected earnings per share and multiples it by the firm's *P/E* ratio to find the value of the firm's stock.

T F 15. Stock market efficiency says that the price of a security accurately reflects all available information about that firm.

T F 16. If a preferred stock is cumulative, then all preferred dividends in arrears must be paid before any dividends can be paid to common stockholders.

T F 17. All common shareholders will always have equal voting rights.

Multiple Choice

1. _________ is the process that links risk and return in order to determine the worth of an asset.
   a. Depreciation
   b. Valuation
   c. Evaluation
   d. Discounting

2. The value of an asset is the _________ of all future cash flows it is expected to provide over a relevant time period.
   a. required return
   b. future value
   c. present value
   d. sum

3. Risk is generally incorporated into the _________ in the present value model.
   a. discount rate
   b. cash flows
   c. timing
   d. total value

4. Stocks are difficult to value because the following information is difficult to accurately obtain.
   a. future dividends
   b. required returns
   c. future growth rate
   d. all of the above

5. The present value of a stock's _________ determines the value of the stock.
   a. dividends
   b. coupon payment
   c. maturity value
   d. coupon payment and maturity value

6. Stock represents
   a. debt.
   b. retained earnings.
   c. equity.
   d. interest.

7. The Gordon growth model requires that analysts determine
   a. a growth rate.
   b. a required return.
   c. next period's dividend.
   d. all of the above.

8. If next period's dividend is $3.00, the required return is 12 percent, and the growth rate is 7 percent, what is the firm's stock price?
   a. $25.00
   b. $60.00
   c. $64.20
   d. $120.00

9. The _________ model is used when valuing preferred stock.
   a. constant growth
   b. zero growth version of Gordon model
   c. Gordon model
   d. book value

10. Hightech Inc. has an expected dividend next year of $3.20 per share, a growth rate for dividends of 8 percent, and a required return of 15 percent. What is the value of a share of Hightech's common stock?
    a. $45.71
    b. $40.00
    c. $21.33
    d. $27.83

11. What is the value of the firm in the event that all assets are sold for their exact accounting value and the proceeds remaining after paying all liabilities are divided among the common stockholders?
    a. Liquidation value
    b. Book value
    c. *P*/*E* multiple
    d. Sell-off value

12. The concept that relies on investors buying or selling any security that they identify as mispriced and thereby causing it price to adjust is known as the
    a. efficient markets theory.
    b. Gordon growth model.
    c. dividend valuation model.
    d. market model.

13. The CURRENT dividend is $2, the growth rate is 5 percent and the required return is 10 percent. What is the value of the stock?
    a. $2.00
    b. $20.00
    c. $40.00
    d. $42.00

14. The financial manager should be concerned about increasing _________ because it will cause the value of the firm's stock to rise.

    a. firm growth.
    b. required return.
    c. risk.
    d. profit.

15. The use of the _________ approach to common stock valuation is superior because it takes into consideration the firm's expected earnings.

    a. book value
    b. present value of interest
    c. liquidation value
    d. *P/E* multiple

16. The key inputs to the valuation process include cash flows (returns), the required return (risk), and

    a. discount rates.
    b. present value.
    c. timing.
    d. expected return.

17. Venture capitalists are responsible for

    a. selling stock to the public.
    b. the IPO.
    c. marketing firm stock.
    d. providing startup capital.

18. Investment bankers are responsible for

    a. providing startup capital.
    b. marketing the firm's stock to the public.
    c. minimizing risk to investors.
    d. providing the firm a fair return.

19. According to the efficient market theory, whenever investors find that the required return of stock is less than the expected return of the stock, the investor will buy the stock. This will

    a. drive the price up.
    b. cause the market to crash.
    c. not affect the price.
    d. drive the price down.

20. The required return for Par Four Inc.'s stock is assumed to be 12 percent, and Par Four has paid the following dividends. What is the value of Par Four's stock under the constant growth model of common stock valuation?

Year	**Dividend per Share ($)**
2013	1.32
2012	1.21
2011	1.16
2010	1.12

a. $11.00
b. $15.11
c. $19.80
d. $21.82

21. American depository shares
a. are backed by American depository receipts.
b. are held in the vaults of banks in the company's home countries.
c. permit foreign investors to buy U.S. companies.
d. trade primarily in the bond markets.

22. Mattel's *Wall Street Journal* quote included a closing value of $26.82 and a net price change of –$0.22. Based on this information, we know that ________ was Mattel's closing price on the prior day.
a. $26.60
b. $26.82
c. $27.04
d. $27.26

23. A U.S. investor can buy
a. American depository receipts (ADRs).
b. American depository shares (ADSs).
c. ADRs and ADSs.
d. neither ADRs or ADSs.

24. Yogi Mat Manufacturing has an outstanding issue of preferred stock with a $30 par value and an 8 percent annual dividend. Suppose Yogi has not paid dividends on its preferred shares in the last year, but investors believe that it will start paying dividends again in one year. What is the value of Yogi preferred stock if investors require a 10 percent rate of return?
a. $48
b. $24
c. $26.18
d. $32.22

25. The common stock of Waters Inc. trades for $35 per share. Investors expect Waters to pay a $1.49 dividend next year, and they expect that dividend to grow at a constant rate forever. If investors require an 11 percent return on this stock, what is the dividend growth rate that they are anticipating?

    a. 6.74 percent
    b. 10.96 percent
    c. 4.26 percent
    d. Cannot be determined from the given information

Essay

1. Distinguish between venture capitalists and investment bankers.

2. What are the sources of error in estimating the value of a share of stock? Which is most likely to be accurate: the computed price of a share of stock or the computed price of a bond?

■ Chapter 7 Answer Sheet

True/False

1. F
2. T
3. T
4. F
5. F
6. T
7. T
8. F
9. T
10. F
11. T
12. T
13. F
14. T
15. T
16. T
17. F

Multiple Choice

1. B
2. C
3. A
4. D
5. A
6. C
7. D
8. B
9. B
10. A
11. B
12. A
13. D
14. A
15. D
16. C
17. D
18. B
19. A
20. D
21. A
22. C
23. B
24. C
25. A

(10.) $$P_0 = \frac{\$3.20}{0.15 - 0.08} = \$45.71$$

(13.) $2 \times (1.05)/(0.1 - 0.05) = 0.42$

(20.)

$PV = -1.12$

$FV = 1.23$

$PMT = 0$

$N = 3$

$CPT\ I = 5.63\%$

$$P_0 = \frac{1.39}{0.12 - 0.0563} = \$21.82$$

(23.)

Div = 30 × 0.08 = 2.40

P = (2.4/0.1) + (2.4/1.1) = 26.18

(PV of perpetuity and PV of div. in arrears)

(25.)

P = D1/(K – g) => g = k – D1/P

g = 0.11– (1.49/35) = 0.0674 = 6.74%

Essay

1. Investment bankers help the firm go public and to market its stock to the public. They operate in the primary market. Venture capitalists provide startup capital to firms that show a promising future.
2. An analyst must estimate the future dividends, the future growth rate, and the riskiness of a firm before applying the Gordon growth model. All of these factors are subject to estimation error and make the estimated price of the stock little more than an educated guess. On the other hand, the inputs to the valuation of a bond are known with some degree of confidence.

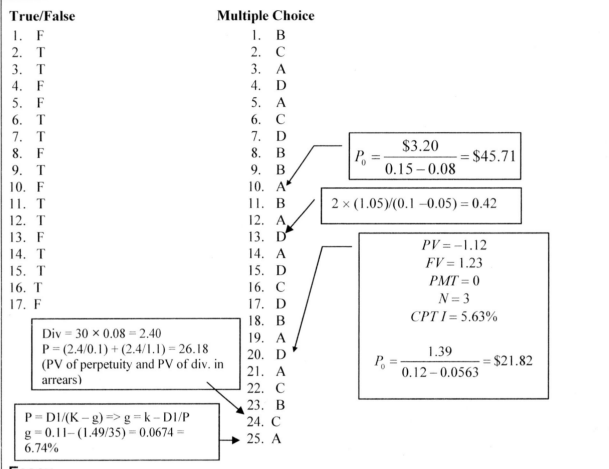

Chapter 8
Risk and Return

■ Chapter Summary

Two of the most important concepts in finance are risk and return. Most of us realize as risk increases, investors will demand a higher return as compensation. What you may not know is that there are different types of risk and only some risk results in increased returns. This chapter explores these concepts and develops the model that relates how much additional return is needed to compensate for added risk.

LG 1 *Understand the meaning and fundamentals of risk, return, and risk preferences.* Most people think of risk as the chance of taking a loss. This is not a sufficient definition of risk in a financial setting. A better definition of risk is the measure of uncertainty surrounding the return an investment will earn. This implies assets can still be risky even when there is no chance of a loss. The total rate of return is the total gain or loss experienced on an investment over a given period. We assume investors are risk averse. This means they avoid risk and when risk increases, investors must be compensated with an increase in return.

LG 2 *Describe procedures for assessing and measuring the risk of a single asset.* If risk is defined as variability in returns, then to measure risk we need a statistic that increases as return variability increases. One such statistic is the standard deviation. An alternative measurement of risk for a single asset is the coefficient of variation. These measures of risk are not appropriate for measuring the risk of an asset held in a portfolio.

LG 3 *Discuss the measurement of return and standard deviation for a portfolio and the concept of correlation.* The return on a portfolio of assets will be the weighted average of the return on the individual assets. The portfolio risk will not be the weighted average of the individual standard deviations because risk will be affected by the relationship between the different assets. The portfolio standard deviation is found by using the formula for the standard deviation of a single asset. Correlation is the statistical relationship between any two series of numbers.

LG 4 *Understand the risk and return characteristics of a portfolio in terms of correlation and diversification and the impact of international assets on a portfolio.* Diversification involves combining assets with low correlation to reduce the risk of the portfolio. The range of risk in a two-asset portfolio depends on the correlation between the two assets. If they are perfectly positively correlated, the portfolio's risk will be between the individual assets' risks. If they are perfectly negatively correlated, the portfolio's risk will be between the risk of the more risky asset and zero.

International diversification can further reduce a portfolio's risk. Foreign assets have the risk of currency fluctuation and political risks.

LG 5 *Review the two types of risk and the deviation and role of beta in measuring the relevant risk of both a security and a portfolio.* In a perfect world, the only relevant risk is nondiversifiable risk because it cannot be eliminated. Diversifiable risk can be eliminated through diversification. Nondiversifiable risk, also called market risk, is measured by beta.

LG 6 *Explain the capital asset pricing model (CAPM), its relationship to the security market line (SML), and the major forces causing shifts in the SML.* The CAPM is a model that projects the required return on an asset given the asset's beta, the return on a risk-free asset, and the performance of the market. The required return is the sum of the return on the risk-free asset plus a risk premium sufficient to compensate an investor for incurring the extra risk of investing in the risky asset.

■ Chapter Notes

Risk and Return Fundamentals

There are two key determinants of share price: risk and return. Risk can be viewed in terms of an asset held in isolation or in terms of an asset held in a portfolio.

Risk Defined

LG 1 Risk is often defined as the chance of a loss. However, in a strict financial context, this is not right. Risk is synonymous with the term *uncertainty*; it is the uncertainty surrounding the return an investment will earn. The more variable the returns to an asset, the more uncertain they are, and the more risky they are.

Return Defined

The total rate of return on an investment is measured as the total gain or loss experienced on behalf of its owner over a given period of time. Return is the change in the asset's value plus any cash distributions received while it is owned. The return is calculated using the following formula.

$$r_t = \frac{P_t - P_{t-1} + C_t}{P_{t-1}}$$

where: r_t = actual, expected, or required rate of return during period t

P_t = price (value) of asset at time t

P_{t-1} = price (value) of asset at time $t - 1$

C_t = cash received from the asset investment in the time period $t - 1$ to t

Risk Preferences

Investors can be either *risk indifferent, risk averse,* or *risk seeking.* A risk indifferent investor does not require any adjustment to the return as risk increases. A risk seeker actually prefers risky investments over less risky investments. The risk-averse investor requires increased returns to compensate for increased risk. We assume that the usual, rational investor is risk averse.

Measuring the Risk of a Single Asset

LG 2 *Scenario analysis*, while not a method for measuring risk, can be used to get a feel for risk. When different scenarios, such as pessimistic (worst case), most likely (expected), and optimistic (best), are evaluated, the greater the difference in the outcomes, the greater is the risk. If changing sales, for example, has little impact on net income, the firm is less risky than one where a small change in sales greatly affects the bottom line. The range is found by subtracting the return associated with the pessimistic outcome from the return associated with the optimistic. It is not unusual to use spreadsheets to create different scenarios as an aid to evaluating the riskiness of an investment.

Probability distributions describe the riskiness of an asset more precisely. A probability distribution is a model that relates probabilities to the associated output. Probability distributions can be shown as either simple bar charts or continuous graphs.

The most common measure of risk is the statistic called the *standard deviation.* The standard deviation measures the dispersion about the mean, or expected return, $\bar{r}$. To compute a standard deviation, the mean is found first. The actual outcomes are subtracted from this mean outcome to find the dispersion. This dispersion is squared to eliminate the negative signs.

$$\bar{r} = \sum_{j=1}^{n} r_j \times Pr_j; \qquad \sigma_r = \sqrt{\sum_{j=1}^{n} \left(r_j - \bar{r}\right)^2 \times Pr_j}$$

where:

r_j = return for the jth outcome

Pr_j = probability of occurrence of the jth outcome

Or when all of the outcomes, r_j, are known *and* their related probabilities are equal, is a simple arithmetic average:

$$\bar{r} = \frac{\sum_{j=1}^{n} r_j}{n} \qquad \sigma_r = \sqrt{\frac{\sum_{j=1}^{n} \left(r_j - \bar{r}\right)^2}{n-1}}$$

where n is the number of observations.

The coefficient of variation is a measure of the relative dispersion. It should be used when comparing the risk of assets that have very different expected returns. The *CV* gives the risk per dollar of return. *CV* is computed by dividing the standard deviation by expected return.

$$CV = \frac{\sigma_r}{\bar{r}}$$

Remember that the standard deviation and the coefficient of variation are useful in measuring the risk of an asset *only when it is held by itself.* Never use these measures when the asset is held as part of a diversified portfolio.

Risk of a Portfolio

LG 3

The risk of any single asset should not be viewed separately from other assets in the portfolio. The investor's goal is to create an efficient portfolio. This is one where the maximum return is achieved with the minimum level of risk. This will only occur in a fully diversified portfolio. Since risk must be viewed along with return, we first show how to compute the return for a portfolio.

Portfolio Return

The return on a portfolio is computed as the weighted average of the returns on each asset in the portfolio.

The standard deviation of a portfolio's return is found by applying the formula for the standard deviation of a single asset, shown above.

Correlation and Diversification

LG 4

Correlation is a statistic that measures the relationship between any two series of numbers. Returns of securities can be either

1. *positively correlated;* if the series move in the same direction;

2. *negatively correlated;* if the series move in opposite directions; or

3. *uncorrelated;* if there is no relationship between the movement of one security and another.

As long as assets are not perfectly positively correlated, diversification can reduce the risk of the portfolio.

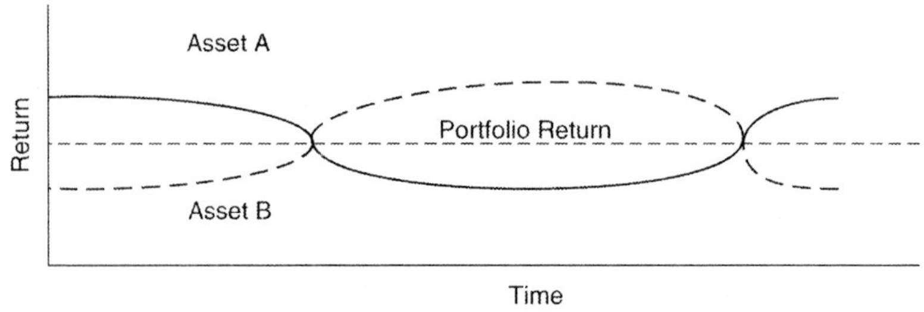

The above graph shows the returns to two perfectly negatively correlated securities. The horizontal dashed line shows the return to the holder of the portfolio. Notice that the portfolio return does not vary over time even though returns to each security do. When assets are perfectly positively correlated, there is no benefit from diversification. In general, the smaller the correlation (the less positive or the more negative), the greater is the benefit of diversification.

International Diversification

Most securities in this country move together. This is because they are all affected by the same factors. For example, when interest rates rise, most firms do poorly because the costs of expansion and debt increase. If the cost of oil were to increase, most firms would suffer. One solution to this problem is to invest in different countries. Because the economies of different countries do not always move together, poor returns from one country may be offset by high returns from another.

There are several types of risk associated with international diversification that should be considered. Currency fluctuations and political turmoil can substantially increase the risk of an international investment.

Diversifiable Risk, Nondiversifiable Risk, and Beta

LG 5

The capital asset pricing model is the theory that links together risk *and* return. To develop this theory, we must first identify the different types of risk.

Total risk can be viewed as consisting of two types: *nondiversifiable risk* and *diversifiable risk.* Diversifiable risk, also called unsystematic risk or firm risk, represents the portion of an asset's risk that is associated with random causes that can be eliminated through diversification. For example, labor problems may affect one firm in a portfolio, but not the others. This may increase the market share of other firms in the portfolio, so the net return to the investor remains unchanged. Nondiversifiable risk cannot be eliminated by diversifying. Factors such as war, inflation, international incidents, and political events tend to affect all firms in similar ways. This kind of risk is also referred to as the risk of the market, or systematic risk.

It does not take a large number of securities to eliminate the bulk of the diversifiable risk. Refer to the following diagram. If an investor holds 15 to 20 securities, most diversifiable risk is eliminated. Because an investor can easily eliminate diversifiable risk, the only risk that is important and the only risk that influences the required return on a security is nondiversifiable or market risk.

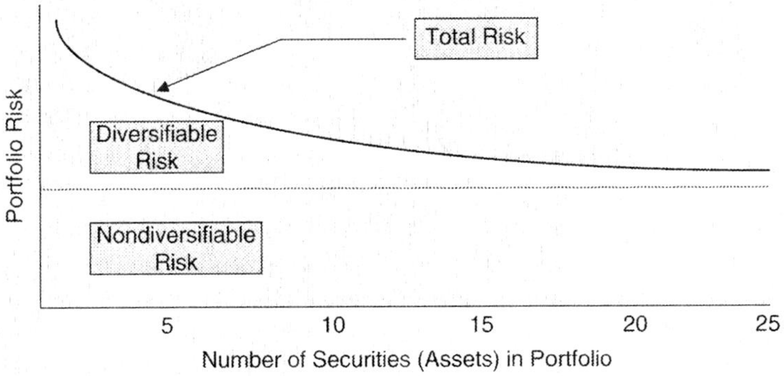

Beta

The beta coefficient is used to measure nondiversifiable risk. It is found by determining how the return on a stock responds to different returns of a portfolio composed of all of the stocks in the whole market.

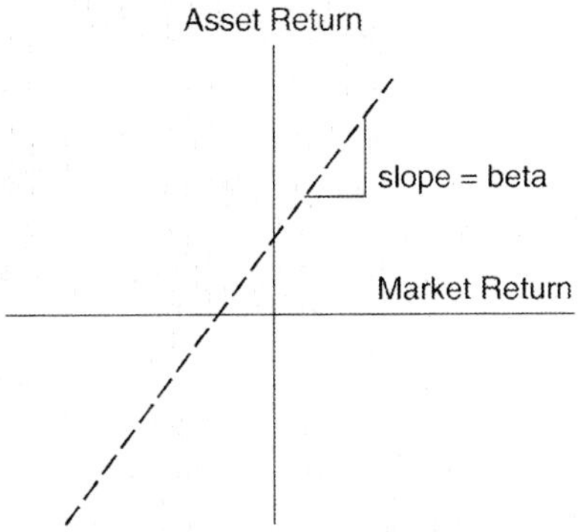

The market returns are on the X-axis, and the asset returns are on the Y-axis. The slope of the line that best fits the plot of these returns is the beta. This line is called the *characteristic line*.

Betas are widely available in many investment guides. The average beta of the whole market is 1. Firms with betas greater than 1 are more risky than the average stock, and firms with betas less than 1 are less risky. For example, most utilities have low betas, say 0.50 to 0.75. Young, high-growth stocks have higher betas, say 1.25 to 2.00.

Portfolio Beta

The beta of a portfolio is found by finding the weighted average of the betas of the firms in the portfolio.

$$b_p = (w_1 \times b_1) + (w_2 \times b_2) + \cdots + (w_n \times b_n)$$

where: b_p = beta of the portfolio
w_1 = proportion of wealth invested in security 1
b_1 = beta of security 1

Compute the portfolio beta in the following example:

Asset	Amount Invested	Weight	Beta	Weight × Beta
IBM	\$25,000	25%	1.15	0.2875
Walmart	\$25,000	25%	1.08	0.27
Apple	\$50,000	50%	1.10	0.55
Portfolio Beta				1.1075 = 1.11

The Model: The Capital Asset Pricing Model (CAPM)

The Equation

LG 6

Using the beta coefficient to measure market risk, the capital asset pricing model is given by the following equation.

$$r_j = R_f + [b_j \times (r_m - R_f)]$$

where: r_j = required return on asset j

R_f = risk-free rate of return

b_j = beta coefficient for asset j

r_m = market return

View the equation as saying the required return on asset j, r_j, is equal to what could be earned with no risk, R_f, plus a risk premium to compensate the investor for choosing a risky investment instead of a risk-free investment. The market risk premium, $(r_m - R_f)$, represents the premium the investor receives for taking the average amount of risk. This average amount is adjusted by multiplying by beta.

Example

What is the required return on a stock with a beta of 1.25, if the return on the market is 0.10 and the risk-free rate is 0.04?

$$r_j = 0.04 + 1.25(0.10 - 0.04) = 0.04 + 0.075 = 0.115$$

The Security Market Line

When the required returns on all stocks are graphed against their corresponding betas, the result is the security market line (SML). The SML will be a straight line.

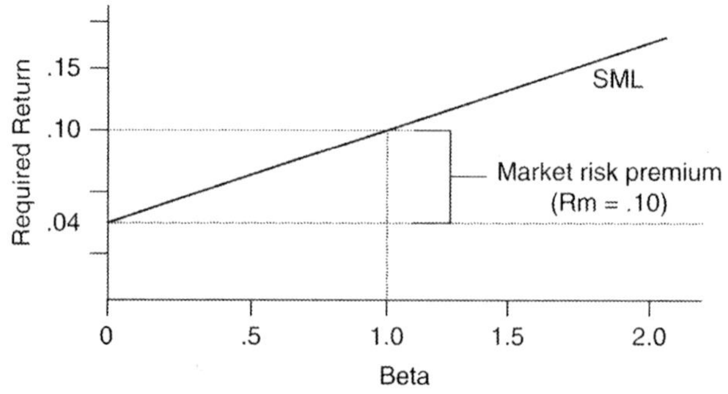

Shifts in the Security Market Line

Market conditions will cause the security market line to shift up or down or to change in slope. For example, if inflation increases, the SML will shift up by the amount of the increase, while maintaining the same slope. If investors become more risk averse, the slope of the SML will increase. This results in offering investors a higher return to compensate them for incurring the risk of the market.

Efficient Markets

Suppose a security offered a return above the required return shown by the SML. As soon as other investors discovered this security, they would rush to purchase it because it represents a "good deal." As many investors attempted to buy, the price of the security would naturally rise. If the price rises, the return that the buyer receives falls. When the price rises enough that the return has fallen to what is offered by other securities of this risk, the buying activity will stop and the security will be in equilibrium.

■ Sample Problems and Solutions

Example 1.

Suppose that you bought IBM stock at $45. Due to the strength of its mainframe business, its stock price rises to $75. Also suppose that $2.00 in dividends are paid. What is your return?

Solution

$$r = \frac{\$75 - \$45 + \$2}{\$45} = 0.7111 = 71.11\%$$

Example 2.

What is the return on the portfolio of securities listed below?

Asset	Amount Invested	Return
IBM	$25,000	0.15
Walmart	$25,000	0.18
Apple	$50,000	0.10

Solution

In the last column the weight (computed by dividing the amount invested in the security by the total amount invested in the portfolio) is multiplied by the return. The product is then summed. The result is the return on the portfolio.

Asset	Amount Invested	Weight	Return	Weight × Return
IBM	$25,000	25%	0.15	0.0375
Walmart	$25,000	25%	0.18	0.0450
Apple	$50,000	50%	0.10	0.05
Return on Portfolio				0.1325

Example 3.

For the stock listed in the following table, calculate the standard deviation. The economy is projected to either boom, be normal, or fall into a recession. Columns 2 and 3 list the probabilities of each of these events along with the stock price under each possibility.

State of Economy	Probability	Stock Price
Boom	0.30	$50
Normal	0.50	20
Depression	0.20	0
$\hat{X}$	—	$25

Solution

State of Economy	Probability	Stock Price	$(\tilde{x}_i - \hat{x})$	$(\tilde{x}_i - \hat{x})^2$	$p_i(\tilde{x}_i - \hat{x})^2$
Boom	0.30	$50	25	625	187.50
Normal	0.50	20	–5	25	12.50
Depression	0.20	0	–25	625	125.00
$\hat{X}$	—	$25	—	—	—

$$\sigma_x = \sqrt{(187.50 + 12.50 + 125.00)} = 18.03$$

First compute the expected value or mean of the stock prices. You should get $25 for the expected value. You next take the outcome for the boom, $50, subtract the expected value ($25), square it, and multiply by the probability of a boom (0.30). You do the same for normal and depression. Finally, sum them up and take the square root.

Example 4.

You are considering an investment portfolio containing two stocks, A and B. Stock A will represent 30% of the dollar value of the portfolio, and stock B will represent the other 70 percent. The expected returns over the next five years, 2015–2019, for each of the stocks are in the table below.

	Expected return	
Year	Stock A	Stock B
2015	12%	17%
2016	14%	15%
2017	14%	13%
2018	15%	11%
2019	15%	9%

a. Calculate the expected portfolio return, r_p, for each of the five years.

b. Calculate the expected value of the portfolio returns, r_p, over the five-year period.

c. Calculate the standard deviation of the expected portfolio returns, SD_p, over the five-year period.

d. How would you characterize the correlation of the return of the two stocks A and B.

e. Discuss any benefits of diversification achieved by the creation of the portfolio.

Solution

a.

Year	Stock A (1)	Stock B (2)	Portfolio Return (1)+(2)
2015	12% × 0.3 = 3.6%	17% × 0.7 = 11.9%	15.5%
2016	14% × 0.3 = 4.2%	15% × 0.7 = 10.5%	14.7%
2017	14% × 0.3 = 4.2%	13% × 0.7 = 9.1%	13.3%
2018	15% × 0.3 = 4.5%	11% × 0.7 = 7.7%	12.2%
2019	15% × 0.3 = 4.5%	9% ×0 .7 = 6.3%	10.8%

b. and c.

$$\bar{r} = \frac{\sum_{j=1}^{n} r_j}{n} \qquad \sigma_r = \sqrt{\frac{\sum_{j=1}^{n}\left(r_j - \bar{r}\right)^2}{n-1}}$$

r_p = (15.5% + 14.7% + 13.3% + 12.2% + 10.8%)/5 = 13.3%

$SD_p = \{ [(15.5 - 13.3)^2 + (14.7 - 13.3)^2 + (13.3 - 13.3)^2 + (12.2 - 13.3)^2 + (10.8 - 13.3)^2]^{(0.5)}\}/4$

$= [14.26^{(0.5)}]/4 = 0.9441$

d. The two stocks, A and B, are negatively correlated because they stay the same or move in the opposite direction each year but not by the same amount so they are not perfectly negatively correlated.

e. When you combine two stocks that are negatively correlated, you reduce the risk of the portfolio.

Example 5.

Review the following and indicate which investment you would choose if you were risk indifferent, a risk seeker, or risk averse. You currently own a firm that earns 13% and has a risk index of 8%.

Investment	Expected Return	Expected Risk Index
Zero Inc.	10%	10%
Batman Enterprises	13%	12%
Spiderman Limited	20%	6%

Solution

a. If you are risk indifferent, you do not care whether an increase in risk is accompanied by an increase in return. In the above example, you would choose Batman Enterprises because as the risk goes up, there is no increase in return.

b. If you are a risk seeker, you want as much risk as possible given your level of return. This would lead you to choose Zero Inc. and Batman Enterprises because they give greater risk with lower return.

c. If you are risk averse (normal), you will choose Spiderman Limited because it gives the highest return for the level of risk.

Example 6.

Compute the coefficient of variation for Bongos Galore if the expected return is 20% and the standard deviation of the returns is 8%.

Solution

The coefficient of variation is a measure of the risk per unit of return earned.

$$CV = 0.08/0.20 = 0.40$$

Example 7.

Given the following returns on the market portfolio and on Tony's Tune-up Shoppe, estimate Tony's beta graphically.

Year	2007	2008	2009	2010	2011	2012	2013
Tony's	5%	8%	10%	6%	4%	–2%	10%
Market	2.5%	4%	5%	3%	2%	–1%	5%

Solution

Begin by graphing the returns with the market return on the *X*-axis and Tony's returns on the *Y*-axis. Using the rise over the run method, estimate the slope of the line drawn that best connects the dots. The slope is the firm's beta. In this case, it is equal to 2. A beta of 2 means that as the market moves by 1%, the return on the stock will move by 2%.

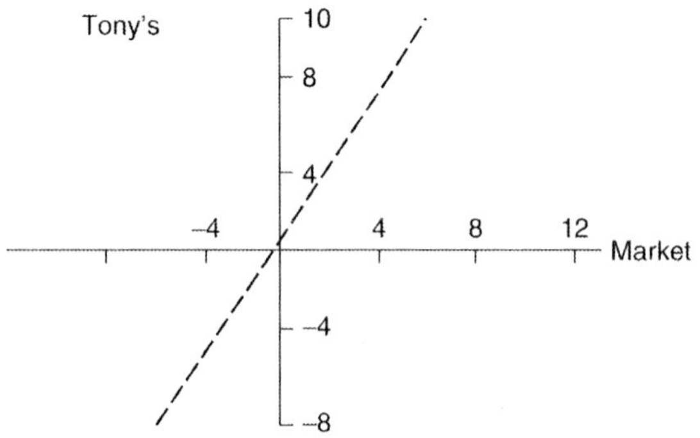

Example 8.

If the risk-free rate of interest is 4percent and the market return is 10 percent, what would be the required return for a security with a beta equal to 2?

Solution

The CAPM is $r_i = R_f + \text{Beta}(r_m - R_f)$. The expected return is then computed as $0.04 + 2(0.1 - 0.04) = 0.16 = 16\%$. The most common mistake made by students using the CAPM is to mix percentages together with decimals in the same equation. Also, make sure that you know where the plus sign is and where the minus sign is.

Example 9.

Benjamin Corporation, a growing computer software developer, wishes to determine the required return on an asset that has a beta of 1.5. The risk-free rate of return is 7% and the expected return on the market is 11 percent.

Solution

$$7\% + 1.5(11\% - 7\%) = 13\%$$

Example 10.

You are interested in investing \$3,000 in asset x with beta of 1.0, \$2,000 in asset y with beta of –0.5, and \$5,000 in asset z with beta of 2.0. What is the beta of the portfolio?

Solution

First compute the asset weights. These are the percentages invested in each of the assets. There is a total of \$10,000 invested in the three assets. Thirty percent is invested in asset x (\$3,000/\$10,000), 20 percent is invested in asset y, and 50 percent is invested in asset z. Now find the weighted average of the betas.

$$B_p = 0.3(1) + 0.2(-0.5) + 0.5(2) = 1.2$$

Example 11.

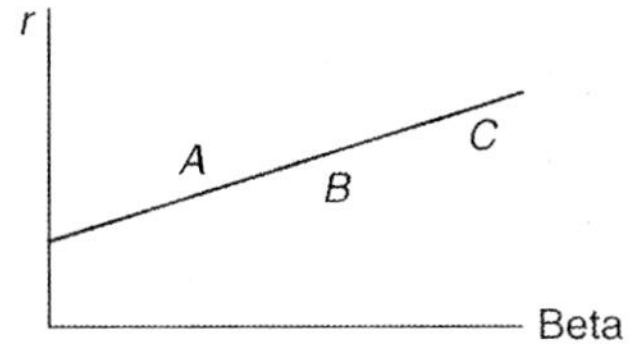

Which of the above assets would you select? Why?

Solution

A is above the security market line. This means that it provides a higher return for its level of risk than do other securities. For this reason you would select it. *B* and *C* are below the security market line and provide a lower return for their level of risk than do other securities. For this reason, reject *B* and *C*.

Example 12.

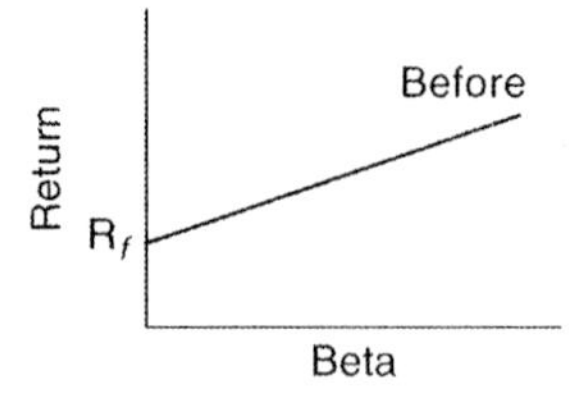

Remember that R_f includes the real rate of interest and inflation and that the market risk premium is a measure of risk aversion. If investors become more risk averse, what happens to the slope of the SML?

Solution

If investors become more risk averse, a higher premium will have to be paid to induce them to invest in risky securities. This implies that the market risk premium increases. Remember that the market risk premium is the slope of the security market line.

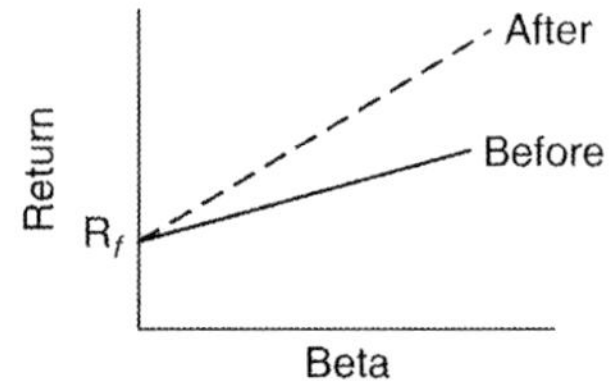

Example 13.

If the inflation premium increases, what happens to the SML?

Solution

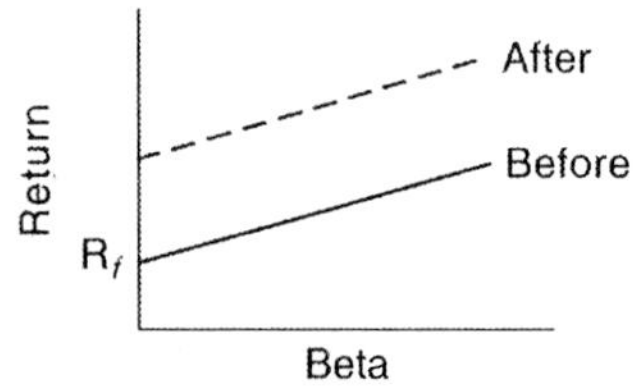

If expected inflation increases, r_m and R_f will both increase by a like amount. Because the slope of the security market line is $(r_m - R_f)$, there will be no change in the slope. The intercept (R_f) will increase.

■ Quick Drill Self Test

The following are a sample of easy, quick practice problems you can use to test your risk and return skills. The answers follow.

1. Compute the holding period return for a security you bought for \$30.00 two years ago and sold last week for \$35.00. You received \$1.00 in dividends while you owned the stock.

2. You believe that next year there is a 30 percent probability of a recession and a 70 percent probability that the economy will be normal. If your stock will yield –10 percent in a recession and 20 percent in a normal year, what is your expected return?

3. What is the standard deviation of the security discussed in Problem 2?

4. If the standard deviation of a security's returns is 14 and the expected return is 11 percent, what is the coefficient of variation?

5. You hold a portfolio composed of 20 percent security A and 80 percent security B. If A has an expected return
of 10 percent and B has an expected return of 15 percent, what is the expected return from your portfolio?

6. Suppose you rebalance the portfolio discussed in Problem 5 such that it is now composed of 80 percent security A and 20 percent security B. If A has an expected return of 10 percent and B has an expected return of 15 percent, what is the expected return from your portfolio?

7. If you were to choose two of the firms listed in Table 7.2 to combine in a portfolio and if your only goal was to minimize risk, which two would you choose?

8. Estimate the beta of a firm assuming that when the market goes up 5 percent this firm's returns rise 20 percent and when the market falls 10 percent this firm's returns fall 40 percent.

9. Estimate the beta of a firm assuming that when the market goes up 5 percent this firm's returns rise 2.5 percent and when the market falls 10 percent this firm's returns fall 5 percent.

10. What is the beta of a portfolio composed of 40 percent security A with a beta of 2 and 60 percent security B with a beta of 0.75?

11. What is the beta of a portfolio composed of 40 percent security A with a beta of 1.5 and 60 percent security B with a beta of –0.25?

12. If the average return on the market is 21 percent and the risk-free security has a return of 6 percent, what is the market risk premium?

13. Assume that the risk-free rate is 5 percent, the return on the market is 15 percent, and that a firm's beta is 0.5. What return must you earn to be satisfied that you are being compensated fairly for the risk of the firm?

14. Using the assumptions made above about the risk-free rate and the return on the market, what will be the required return for a security with a beta of 1.5?

15. You have analyzed the firm discussed in Problem 14 and determined that its expected return is 19.5 percent. Would you choose to buy it?

16. If investors in general agree that the market is less likely to post a loss, are market risk premiums likely to rise or fall?

17. The risk-free rate of interest is 5 percent. The market risk premium falls from 10 percent to 8 percent. How much does the required return fall for a security with a beta of 1.5?

18. The risk-free rate of interest is 5 percent. The market risk premium falls from 10 percent to 8 percent. How much does the required return fall for a security with a beta of 0.75?

19. You have become convinced that inflation has been tamed. You have been using a real rate of 2% and an inflation rate of 4% to estimate the risk-free rate to be 6 percent. The market rate is also estimated using an assumed inflation rate of 4 percent. If you believe that inflation will be 3 percent in the future, how much will the required return on your portfolio change by if the return on the market is 10 percent and your portfolio's beta is 1.2? (Note that both the market rate and the risk-free rate will change.)

20. Refer to the data in Problem 19. How much will the required return on your portfolio change by if your portfolio's beta is 0.5? (Assume inflation is 4 percent.)

Answers to Quick Drill Self Test

1. 20%
2. 11%
3. 0.1374
4. 1.27
5. 14%
6. 11%
7. Fairchild and SWA
8. 4
9. 0.5
10. 1.25
11. 0.45
12. 15%
13. 10%
14. 20%
15. No
16. Fall
17. 3%
18. 1.5%
19. 1%
20. 2.8%

■ Study Tips

1. Remember that the firm's standard deviation and coefficient of variation measure total risk are not appropriate measures of risk of securities held in a portfolio. In a portfolio, the only risk that matters is market risk, measured by beta, because the rest is eliminated by diversification.

2. The CAPM provides a way of calculating the required return on a security. The required return is how much you must earn to feel fully compensated for holding a risky investment.

3. Be able to distinguish between the characteristic line and the security market line. The characteristic line relates one security to the market. The slope of the characteristic line is beta. The security market line relates the required return to risk. The slope of the security market line is the market risk premium.

4. Be able to distinguish between diversifiable and nondiversifiable risk.

■ Student Notes

____

____

____

____

____

■ Sample Exam—Chapter 8

True/False

T F 1. Most investors are risk averse because for a given increase in return they require an increase in risk.

T F 2. The coefficient of variation is useful for evaluating the risk of a security held in a portfolio.

T F 3. Longer-lived assets usually are more risky because it is more difficult to accurately forecast cash flows that occur far into the future.

T F 4. Adding an asset to a portfolio that is perfectly positively correlated with existing portfolio returns will have no affect on portfolio risk.

T F 5. There is no possible way that combining assets into a portfolio will result in a portfolio that is riskier than its riskiest asset.

T F 6. Unless assets are negatively correlated, combining assets into a portfolio will not reduce portfolio risk.

T F 7. A firm with a negative beta may have high sales when the economy weakens and low sales when the economy recovers.

T F 8. Diversifiable risk is the only risk that influences the required return because nondiversifiable risk can be eliminated.

T F 9. Beta measures nondiversifiable, or market, risk.

T F 10. Beta is the slope of the security market line.

T F 11. The characteristic line maps the relationship between the return on the market and the return on a security. Its slope is beta.

T F 12. As investors become more risk averse, the market risk premium increases and the security market line becomes steeper.

T F 13. Fixed income securities, such as bonds or preferred stocks, usually have betas equal to zero.

T F 14. Risk can best be defined in finance as the chance of a loss.

T F 15. The standard deviation is a measure of relative dispersion and is used in comparing the risk of assets with differing expected returns.

T F 16. Standard deviation can be used to measure risk on an ex-ante basis using historical returns.

T F 17. The beta of a risk-free asset is –1.

T F 18. The beta of the market portfolio is 1.

Multiple Choice

1. __________ is the variability in returns associated with an investment.
   a. Median
   b. Risk
   c. Coefficient of variation
   d. Profit

2. You compute the __________ on an asset by subtracting the beginning price from the ending price, adding any cash distributions, and dividing by the beginning price.
   a. return
   b. risk
   c. profit
   d. standard deviation

3. A person who prefers more risk to less risk and who does not require additional return to compensate for increased risk is
   a. risk averse.
   b. risk neutral.
   c. risk seeker.
   d. smart.

4. Assume that last year you purchased 100 shares of Rubber Ducky Inc. for $25 per share. The price 1 year later is $27 per share. If a $2.50 dividend was paid, what is the return?
   a. $50
   b. 8 percent
   c. 2.50 percent
   d. 18 percent

5. Assume that you purchased stock last year for $27. Today's price is $25. What is the return?
   a. $2.00
   b. 8 percent
   c. –7.41 percent
   d. $2.00

6. One method to evaluate the riskiness of an investment opportunity is to estimate the cash flows under the most optimistic, most likely, and most pessimistic set of assumptions. This type of analysis is called
   a. break-even analysis.
   b. marginal analysis.
   c. scenario analysis.
   d. probability analysis.

7. A measure of the dispersion of returns around the mean return is the
   a. probability distribution.
   b. standard deviation.
   c. coefficient of variation.
   d. student's *t*.

8. Given the following expected returns and standard deviations of assets A, B, and C, which asset should the prudent financial manager select?

Asset	Expected Return	Standard Deviation
A	10 percent	10 percent
B	12 percent	10 percent
C	12 percent	8 percent

   a. A
   b. B
   c. C
   d. Impossible to tell

9. What is the coefficient of variation for asset A if the standard deviation is 0.02, the expected return is 0.10, and the probability of a return equal to 0.10 is 25 percent?
   a. 0.02
   b. 0.20
   c. 0.05
   d. 5

10. A portfolio that provides the maximum return for a given level of risk is known as the
   a. optimal portfolio.
   b. risk-averse portfolio.
   c. continuous portfolio.
   d. efficient portfolio.

11. If the returns on one asset move in the same direction as the returns on another asset, then the assets are
   a. positively correlated.
   c. not correlated.
   b. negatively correlated.
   d. perfectly positively correlated.

12. If you put $26,000 in asset A, $33,000 in asset B, and $55,000 in asset C, and if the expected return to A is 10 percent, to B is 12 percent, and to C is 14 percent, what is the expected return on the portfolio?
    a. 12 percent
    b. 0.125
    c. 0.228
    d. 0.1425

Use the following table to answer Questions 13–16.

Asset	Return	Beta	Proportion
A	10%	0.75	0.20
B	12%	1.00	0.40
C	14%	1.25	0.40

13. What is the expected return for the portfolio?
    a. 0.11
    b. 0.124
    c. 0.37
    d. 0.4

14. The portfolio beta is
    a. 0.75.
    b. 1.
    c. 1.05.
    d. 1.25.

15. Referring to the above table, which firm is most risky in a portfolio context?
    a. A
    b. B
    c. C
    d. Impossible to tell

16. Referring to the above table, which firm is most risky if held in isolation (i.e., not part of a portfolio)?
    a. A
    b. B
    c. C
    d. Impossible to tell

17. Combining several assets into a portfolio so that the risk of the portfolio is reduced is known as
    a. valuation.
    b. diversification.
    c. liquidation.
    d. risk aversion.

18. Another term for nondiversifiable risk is
    a. systematic risk.
    b. nonsystematic risk.
    c. firm specific risk.
    d. both b and c.

19. The beta of the market portfolio is
    a. 0.
    b. 1.
    c. 1.5.
    d. variable according to the level of risk aversion.

20. The slope of the security market line increases when
    a. inflation increases.
    b. inflation decreases.
    c. the risk premium increases.
    d. investors become less risk averse.

21. When the capital asset pricing model is graphed, the resulting line is called the
    a. characteristic line.
    b. security market line.
    c. standard deviation.
    d. beta.

22. If the beta of an asset is 1.25, the risk-free rate of interest is 0.04, and the return on the market portfolio is 0.12, what is the required return on the asset?
    a. 0.12
    b. 0.14
    c. 0.16
    d. 0.24

23. What is the market risk premium for the asset discussed in Problem 22 above?
    a. 1.25
    b. 0.12
    c. 0.08
    d. 0.16

24. If the required return on a security is 10 percent, the risk-free rate is 0.05, and the return on the market is 0.14, what is the beta of the firm?
    a. 0.14
    b. 1
    c. 1.4
    d. 0.56

25. As risk aversion decreases, due to greater confidence in the long-term health of the economy,
    a. firms' betas will increase.
    b. investors' required return will decrease.
    c. investors' required return will increase.
    d. firms' betas will decrease.

26. _________ analysis is a technique for assessing risk that uses several possible solutions in order to obtain a sense of the variability among returns.
    a. Monte Carlo
    b. Range
    c. Scenario
    d. Situation

27. Actual risk may be _________ the expected risk.
    a. greater than
    b. equal to
    c. less than
    d. All of the above

28. Vy Le will only accept companies that have a coefficient of variation of 0.82 or better. Which of the following companies is acceptable?
    a. AXEL: Return = 10 percent, Standard Deviation = 10 percent
    b. BENRIL: Return = 12 percent, Standard Deviation = 10 percent
    c. CAMPEN: Return = 8 percent, Standard Deviation = 10 percent
    d. DEMPSZY: Return = 14 percent, Standard Deviation = 10 percent

29. If an asset has a beta of 1.1 and the market return increased by 12 percent, what impact would this change have on the asset's return?
    a. The asset's return would be expected to be 12 percent
    b. The asset's return would be expected to be 13.2 percent.
    c. The asset's return would be expected to be 11.1 percent
    d. The asset's return would be expected to be 1.2 percent.

Use the table below for question 30:

Asset	Beta
A	0.40
B	2.10
C	–0.30
D	0.95

30. If you believed the market return would go up in the near future you would invest in Asset _____, but if you believed the market return would go down in the near future you would invest in Asset _____.
    a. Asset B; Asset A
    b. Asset C; Asset B
    c. Asset B; Asset C
    d. Cannot determine from the information given.

Essay

1. Distinguish between an investor's expected rate of return and the required rate of return. How is each computed? If the expected rate of return is 12 percent and the required rate of return is 10 percent, will investors buy the security?

2. Given the following information, should the asset be purchased?

State of the Economy	**Probability**	**Return**
Boom	0.25	0.16
Normal	0.5	0.14
Bust	0.25	0.12

The risk-free rate of interest is 5 percent, the return on the market is 12 percent, and the beta of the firm is 1.0.

3. Discuss some of the decisions that need to be made when calculating beta and how these decisions can lead to a variety of different betas for the same asset.

■ Chapter 8 Answer Sheet

True/False

1. F
2. F
3. T
4. T
5. T
6. F
7. T
8. F
9. T
10. F
11. T
12. T
13. T
14. F
15. F
16. T
17. F
18. T

Multiple Choice

1. B
2. A
3. C
4. D
5. C
6. C
7. B
8. C
9. B, $0.02/0.10 = 0.2$
10. D
11. A
12. B, $\$26,000/\$114,000 = 0.228$
    $0.228(0.10) + 0.289(0.12) + 0.482(0.14) = 0.125$
13. B, $0.1(0.2) + 0.12(0.4) + 0.14(0.4) = 0.124$
14. C, $0.2(0.75) + 0.4(1) + 0.4(1.25) = 1.05$
15. C
16. D, (need to see either SD or CV)
17. B
18. A
19. B
20. C
21. B
22. B, $0.04 + 1.25(0.12 - 0.04) = 0.14$
23. C, $0.12 - 0.04 = 0.08$
24. D, $0.10 = 0.05 + B(0.14 - 0.05)$, $B = 0.55$
25. B
26. C
27. D
28. D
29. B $1.1 \times 12\% = 13.2\%$
30. C (want the highest positive beta in increasing markets and the most negative beta in decreasing markets)

Essay

1. The expected return is what you think you will earn based on the different possible states of the economy and their associated probabilities. This is computed by finding the weighted average of the returns earned in each of the states of the economy. The required return is the return needed to fully compensate the investor for investing in a risky asset. It is computed by using the CAPM. If the expected return is greater than the required return, the investor will buy the security.

2. To solve this problem you must first compute the expected return. This expected return can then be compared to the required return computed from the CAPM.

   $Exp(R) = 0.25(0.16) + 0.5(0.14) + 0.25(0.12) = 0.14$

   $Req(R) = 0.05 + 1(0.12 - 0.05) = 0.12$

   Because the required return is less than the expected return, the security should be purchased.

3. Some of the decisions that need to be made when calculating beta and that can lead to different betas for the same asset are:

   a) The time horizon: should you be measuring beta over a seven-year period, a 10-year period, a 25-year period, etc.

   b) Should you use daily returns, weekly returns, quarterly returns, or annual returns?

   c) What should you use as the market portfolio? Should it be the S&P 500, the Russell 2000, the Dow Jones Industrial? Should there be corporate bonds, municipal bonds, government bond, money market securities, or real estate in the market portfolio?

   As you can imagine, every time one of these factors changes, you will get a different beta. So although the text gives you a beta to use in the problems, you should be aware of some of the different factors that affect betas and therefore the required returns.

Chapter 9
The Cost of Capital

■ Chapter Summary

This chapter covers how firms estimate their cost of capital, the rate that will be used in the techniques in the next two chapters to determine if a company should accept future investments. The most difficult part of this chapter is learning how to account for changing discount rates that result from increasing the demand for funds by firms that have many investment opportunities. In this chapter it is assumed that the optimal or target mix of debt and equity is known. We conclude the long-term financing discussion in a later chapter by discussing the determinants of this mix.

LG 1 *Understand the basic concept and sources of capital associated with the cost of capital.* The cost of capital represents the firm's cost of financing, and is the minimum rate of return a project must earn to increase firm value. There are four main sources of capital discussed in this chapter. Debt is the lowest-cost source of capital because interest paid to bondholders is tax deductible to the firm. Equity—either common or preferred stock—can be sold to raise capital. In addition, equity from common stockholders can be raised from retaining earnings rather than paying dividends.

LG 2 *Explain what is meant by marginal cost of capital.* The relevant cost of capital for a firm is the marginal cost of capital necessary to raise the next marginal dollar of financing the firm's future investment opportunities. A firm's future investment opportunities in expectation will be required to exceed the firm's cost of capital.

LG 3 *Determine the cost of long-term debt, and explain why the after-tax cost of debt is relevant cost of debt.* In Chapter 6 you learned to compute the yield to maturity for bonds. This method is used to compute the before-tax cost of debt. The after-tax cost of debt is calculated by multiplying the before-tax cost of debt by 1 minus the tax rate. Because interest payments are tax deductible, only after-tax costs are used.

LG 4 *Determine the cost of preferred stock.* The cost of preferred stock is the ratio of the preferred stock dividend to the firm's net proceeds from the sale of preferred stock. The preferred stock dividend that must be paid is not tax deductible; therefore no tax adjustment is made to the cost.

LG 5 *Calculate the cost of common stock equity, and convert it into the cost of retained earnings and the cost of new issues of common stock.* There are two models introduced in Chapters 7 and 8 that can be used to compute the cost of equity financing (the constant-growth model and the CAPM). The reason for using multiple models is because the inputs to each model are uncertain. By combining the results, a better estimate of cost can be achieved.

LG 6 *Calculate the weighted average cost of capital (WACC) and discuss alternative weighting schemes.* Once all of the component costs are found, the weighted average cost of capital is computed where the weights are the proportion of each source of capital in the firm's target capital structure. This WACC is used as the discount rate in capital budgeting problems.

■ Chapter Notes

An Overview of the Cost of Capital

LG 1

The cost of capital is the minimum rate of return a firm must earn on an investment to grow firm value. A weighted average cost of capital should be used to find the expected average future cost of funds over the long run. It is the average cost of all the sources of financing for the firm. The cost of capital plays a major role in the long-term investment decisions of the firm because it has a substantial part in determining the validity of a project. Selection of projects with returns in excess of the cost of capital increases firm value.

The Basic Concept

Capital structure consists of the long-term sources of financing a firm chooses to use. There are four basic sources of long-term funds available to the firm. These are long-term debt, preferred stock, retained earnings, and common stock. Long-term debt reflects money borrowed from creditors, whereas preferred stock, common stock, and retained earnings are classified as equity. The firm should try to maintain an optimal mix of debt and equity financing. This optimal mix is known as the target capital structure and will be discussed in the next chapter. In practice, firms attempt to keep financing within a target range, such as 40% to 50% debt.

Marginal Cost of Capital

LG 2

The relevant cost of capital for a firm is the marginal cost of capital necessary to raise the next marginal dollar of financing the firm's future investment opportunities. A firm's future investment opportunities will be required to exceed the firm's cost of capital.

The Cost of Long-Term Debt

LG 3

The cost of long-term debt, r_i, is the after-tax cost today of raising long-term funds through borrowing. In most cases, long-term debt is borrowed in the form of bonds that pay interest semiannually. For simplicity and ease of calculation, we will assume the interest is paid annually. The net proceeds of these bonds consist of the selling price of the bond minus the flotation costs, which are the total costs of issuing and selling a bond.

Before-Tax Cost of Debt

The before-tax cost of debt, r_d, for a bond can be found in one of three ways: (1) quotation, (2) calculation, and (3) approximation.

When the net proceeds from the sale of a bond are equal to its face value, the coupon interest rate can be used for the bond's before-tax cost. Also, when the yield to maturity (YTM) for other bonds of similar risk is available, the similar bond's YTM can be used for the firm's before-tax cost of long-term debt. This is termed a cost quotation.

When the net proceeds from the sale of the bond differ from the face value of the bond, it is necessary to either calculate the cost or approximate the cost. The calculation of the before-tax cost of debt for a bond involves finding the YTM for the bond either by using a financial calculator or an electronic spreadsheet.

The equation used for approximating the before-tax cost of long-term debt for a bond is demonstrated in the following formula:

$$r_d = \frac{I + \frac{M - N_d}{n}}{\frac{N_d + M}{2}}$$

where: I = annual interest in dollars

M = par value of bond

N_d = net proceeds from the sale of debt (bond)

n = number of years to the bond's maturity

After-Tax Cost of Debt

As mentioned earlier, the cost of financing must be stated on an after-tax basis. The interest on debt is tax deductible, so it reduces the firm's taxable income by the amount of deductible interest. The equation used to find the after-tax cost of debt is demonstrated in the following formula:

$$r_i = r_d \times (1 - T)$$

where: r_i = after-tax cost of debt

r_d = before-tax cost of debt

T = corporate tax rate

Example

Velvet Corporation has a before-tax cost of debt of 8.4% and a corporate tax rate of 40%. What is the after-tax cost of debt for Velvet Corporation?

$$r_i = 0.084 \times (1 - 0.40)$$

$$r_i = 5.04\%$$

Because the interest on the debt is tax deductible, Velvet is paying only 5.04% for the debt and is receiving a 3.36% (or 8.4 × 0.4) tax deduction for every dollar spent on interest.

The Cost of Preferred Stock

LG 4

Preferred stock is a special form of ownership because its dividends must be distributed before any dividends are paid to the common stockholders. Most preferred stock dividends are stated as a dollar amount, but sometimes they are stated as a percentage of the stock's par value. In order to calculate the cost of preferred stock, it is necessary to convert the dividend of any preferred stock stated as a percentage into a dollar amount. Once the dividend is stated as a dollar amount, the cost of the preferred stock can be found by using the following equation:

$$r_p = \frac{D_p}{N_p}$$

where: r_p = cost of preferred stock

D_p = annual dollar dividend

N_p = net proceeds from the sale of the stock

Preferred stock dividends are paid out of the firm's after-tax cash flows, so no tax adjustment is necessary.

The Cost of Common Stock

LG 5

There are two forms of common stock financing: retained earnings and new issues of common stock. Before finding each of these costs, the cost of common stock equity must be estimated. The cost of common stock equity, r_s, is the rate at which investors discount the expected dividends of the firm to determine its share value. There are two techniques that are used to estimate r_s. The constant-growth valuation model, also known as the Gordon model, was discussed in Chapter 7 and is presented in the following equation:

$$P_0 = \frac{D_1}{r_s - g}$$

where: P_0 = value of common stock

D_1 = per-share dividend expected at the end of year 1

r_s = required return on common stock

g = constant rate of growth in dividends

To find the cost of common stock equity, simply solve the equation for r_s.

$$r_s = \frac{D_1}{P_0} + g$$

Example

Velvet Corporation wishes to estimate its cost of common stock equity, r_s. The firm expects to pay a dividend, D_1, of \$2.87 at the end of 2014. The current market price, P_0, of its common stock is \$35, and the dividends paid on its common stock over the past five years are presented in the following table:

Year	Dividend
2013	\$2.70
2012	2.56
2011	2.42
2010	2.26
2009	2.14

Using a financial calculator, the growth rate has been determined to be 6 percent. (See Chapter 5 to review growth rates.) Substituting the information into the equation used to solve for r_s will result in:

$$r_s = \frac{\$2.87}{\$35} + 6\% = 8.2\% + 6\% = 14.2\%$$

The 14.2 percent represents the return required by *existing* shareholders on their investment.

The capital asset pricing model (CAPM) introduced in Chapter 8 can also be used to estimate the cost of common stock equity. The CAPM describes the relationship between the required return, or the cost of common equity in this case, and the nondiversifiable risk of the firm as measured by the beta coefficient, β.

The following equation represents the basic CAPM:

$$r_s = R_F + [b \times (r_m - R_F)]$$

where:

R_F = risk-free rate of return

b = beta coefficient

r_m = market return; return on the market portfolio of assets

It is important to note the differences between the constant-growth valuation model and the CAPM in finding the cost of common stock. The CAPM technique directly considers the firm's risk, through its inclusion of a beta term, in determining the required rate of return on common stockholders. By contrast, the constant-growth model uses the market price in the denominator. This price is an indication of the expectations of investors in the marketplace regarding risk and return.

The Cost of Retained Earnings

Earnings, after all expenses and preferred dividends are paid, belong to the common stockholders. The firm can pay out all earnings or pay only a portion of the earnings as dividends and retain the balance. The earnings retained represent an additional investment by the existing common stockholders. Viewing retained earnings as the equivalent of the issuance of additional common stock, the cost of retained earnings, r_r, can be set equal to the cost of common stock equity.

$$r_r = r_s$$

The Cost of New Issues of Common Stock

The cost of new issues of common stock is greater than the cost of retained earnings because of underpricing and flotation costs. The cost of new issues of common stock is found by calculating the cost of common stock equity after the amount of underpricing and flotation costs are considered. To sell a new issue of common stock, it normally has to be underpriced, which means selling the stock at a price below the current market, P_0. In addition, flotation costs must be paid to the investment banker for selling the new shares. Both underpricing and flotation costs will reduce the proceeds from the sale of new common stock. The cost of new issues can be found by using the constant-growth valuation model used for finding the cost of existing common stock. Simply adjust the current market price of the stock, P_0, for any underpricing and flotation cost.

$$r_n = \frac{D_1}{N_n} + g$$

Where:

r_n = cost of new issues

N_n = net proceeds from the sale of new issues

The Weighted Average Cost of Capital

LG 6

The weighted average cost of capital (WACC) is found by weighting the cost of each specific type of capital by its proportion in the firm's capital structure to find the average cost of funds over the long run. We previously discussed the procedure for finding the specific after-tax cost of each source of capital. To determine the weights to use, simply add the total dollar amount of long-term financed capital and divide each component by the total amount to find its proportion.

$$r_a = (w_i \times r_i) + (w_p \times r_p) + (w_s \times r_{r\ or\ n})$$

where:

w_i = proportion of long-term debt in capital structure

w_p = proportion of preferred stock in capital structure

w_s = proportion of common stock equity in capital structure

$w_i + w_p + w_s = 1.0$

Example

Velvet Corporation has the following capital structure:

Source	Amount
Long-term debt	$200,000
Preferred stock	25,000
Common stock equity	275,000
Total	$500,000

The total dollar amount of financed capital is $500,000. To determine each component's weight, simply divide the amount of the financed capital being used by $500,000. For example, to get the weight of long-term debt, divide $200,000 by $500,000.

$$w_i = \frac{\$200,000}{\$500,000} = 0.40$$

$$w_p = \frac{\$\ 25,000}{\$500,000} = 0.05$$

$$w_s = \frac{\$275,000}{\$500,000} = 0.55$$

Remember that these weights should total to 1. Once the weights have been determined, the cost of each component can be multiplied by its weight to determine the WACC.

Where:

w_i = proportion of long-term debt in capital structure

w_p = proportion of preferred stock in capital structure

w_s = proportion of common stock in capital structure

Using *target capital structure weights*, the firm is trying to develop a capital structure that is optimal for the future, given present investor attitudes toward financial risk. Target capital structure weights are most often based on desired changes in historical book value weights. Unless significant changes are implied by the target capital structure weights, little difference in the weighted marginal cost of capital results from their use.

Sample Problems and Solutions

Example 1. Cost of Debt

Find the cost of debt for a bond that is selling for \$950, has five years to maturity, and has a coupon interest rate of 10 percent. Assume the tax rate is 40 percent.

a. Equation Solution

The cost of debt is found using the procedure shown earlier for computing the yield to maturity. The difference is that the market price is adjusted to reflect any discounts, fees, or costs that the firm must pay. In this problem, no costs are mentioned, so we assume that there are not any. The YTM can be found easily on a financial calculator or by using the approximation equation.

$$\text{Approximate Yield (YTM)} = \frac{I + \frac{M - N_d}{n}}{\frac{N_d + M}{2}}$$

$$\text{YTM} = \frac{90 + \frac{(1000 - 950)}{5}}{\frac{1000 + 950}{2}} = 0.113$$

$$r_d = 0.113 \times (1 - 0.4) = 0.0678 = 6.78\%$$

b. Calculator Solution

$N = 5$

$PV = 950$

$PMT = -100$

$FV = -1{,}000$

$CPT\ I = 11.37$

$r_d = 0.1137 \times (1 - 0.4) = 0.0682 = 6.82\%$

Example 2. Cost of Long-Term Debt

Laurie's Rose Distributors is contemplating expanding by selling bonds. Her investment banker advises that the bonds should sell for \$1,100, less 3 percent for fees. If the bonds have a coupon rate of 10 percent and mature in 10 years, what is Laurie's cost of debt? Assume a 40 percent tax rate.

a. Equation Solution

The above estimation equation can be used.

The market price will be \$1,100 – (1,100 × 0.03) = \$1,067.

$$\text{YTM} = \frac{100 + \frac{(1000 - 1{,}067)}{10}}{\frac{1000 + 1{,}067}{2}} = 0.0903$$

$$r_d = 0.0903 \times (1 - 0.4) = 0.0542 = 5.42\%$$

b. Calculator Solution

$N = 10$

$PV = 1{,}067$

$PMT = -100$

$FV = -1{,}000$

$CPT\ I = 8.96$

$r_d = 0.0896 \times (1 - 0.4) = 0.0538 = 5.38\%$

Example 3. Cost of Preferred Stock

Compute the cost of preferred stock financing if the promised annual dividend is \$2.00 and the current market price of the stock is \$22.00.

Solution

Preferred stock pays a dividend in perpetuity, meaning forever. By rearranging the equation we learned for valuing a perpetuity, we can find the return. No tax adjustment is needed because preferred stock dividends are not tax deductible. Because this is not a new issue of preferred stock, there are no flotation costs to be considered.

$$r_p = \frac{D_p}{N_p}$$

$$r_p = D_p/N_p = \$2/\$22 = 0.0909 = 9.09\%$$

Example 4. Cost of Equity

Compute the cost of equity if the common stock is selling for \$35, the growth rate is expected to be 5 percent, and next year's dividend is expected to be \$3.

Solution

Given the available information, the only model that can be used is the Gordon growth model. By rearranging the terms, we find that:

$$r_s = D_1/P + g$$

so

$$r_s = \$3/\$35 + 0.05 = 0.136 = 13.6\%$$

Note: The most common mistake made when using this equation is to mix decimals and percentages. Do not say the $r_s = 3/35 + 5\% = 0.086 + 5 = 5.086\%$.

Example 5. Cost of Equity

Compute the cost of equity if the firm's beta is 1.12, the risk-free rate of interest is 7 percent, and the return on the market is 13 percent.

Solution

The only method to solve for the cost of equity that is possible, given the above information, is the CAPM.

$$r_s = 0.07 + 1.12(0.13 - 0.07) = 0.137 = 13.7\%$$

Example 6. Cost of Equity

Compute the cost of equity if new stock must be sold and the flotation costs are expected to be 5 percent of the current market price of the stock. The stock is selling for \$35, the growth rate is expected to be 5 percent, and the next dividend is expected to be \$3.

Solution

When flotation costs exist, the cost of common stock to the firm rises because the firm receives a lesser amount. In this example, when stock is sold, the public will pay \$35 per share, but the firm will receive only \$33.25 [or, 35 × (1 – 0.05)] of the proceeds.

$$r_s = \$3/\$33.25 + 0.05 = 0.1402 = 14.02\%$$

Note: Compare this result to Example 4. The only difference was that flotation costs are included here. As a result, the cost of equity increases.

Example 7. Weighted Average Cost of Capital

Compute the weighted average cost of capital for Bob's Crystal Pool Supply given that $r_d = 10\%$ (after-tax), $r_p = 12\%$, and $r_s = 13.7\%$. The target capital structure is 60% common stock, 30% debt, and 10% preferred stock.

Solution

$$\text{WACC} = 0.60(0.137) + 0.30\ (0.10) + 0.1(0.12) = 0.1242 = 12.42\%$$

Example 8. Weighted Average Cost of Capital Including Calculation of Individual Costs

Milton Inc. has a target capital structure consisting of 40 percent debt, 10 percent preferred stock, and 50 percent common equity. Its bonds have an 11 percent coupon, paid semiannually, a current maturity of 20 years, sell for \$1,020, and a 5% flotation cost. The firm could sell a \$100 par value preferred stock that pays a 12% annual dividend for \$100, but flotation costs of 5 percent would be incurred. Milton's beta is 1.1, the risk-free rate is 4 percent, and the market risk premium is 5 percent. Milton Inc. expects a constant growth rate of 8 percent and just paid a dividend of \$2.00. Its common stock sells for \$27.00 per share. Flotation costs on new common stock total 10 percent, and the firm's marginal tax rate is 40 percent. Assume the firm has insufficient retained earnings to fund the equity portion of its capital budget. What is the weighted average cost of capital for Milton Inc.?

Solution

You must first find the individual costs of capital:

The before-tax cost of debt: PV= –(1,020 + 50) = –1,070; FV = 1,000; PMT = 110/2 = 55; N = 2 × 20 = 40; CPT I = 5.09, which is a semiannual rate so multiple by 2 => 10.18%.

The after-tax cost of debt: before-tax cost of debt × (1 – tax rate) = 10.18% × (1 – 0.4) = 6.11% = r_d

The cost of preferred: $r_p = D_p/N_p$ = 12/95= 12.63%.

Because the firm has insufficient retained earning, you need to find the cost of new equity.

Cost of new equity: $r_s = D_1/N_p + g = 2/24.3 + 0.08 = 0.0823 + 0.08 = 0.1623 = 16.23\%$

WACC: $r_a = (w_i \times r_i) + (w_p \times r_p) + (w_s \times r_{r\ or\ n})$

WACC = r_a = (0.4 × 6.11%) + (0.1 × 12.63%) + (0.5 × 16.23) = 2.44 + 1.26 + 8.12 = 11.82%

■ Quick Drill Self Test

1. The current market price of a $1,000 par bond is $925. Its semiannual coupon rate is 7 percent, and it matures in five years. What is the pre-tax cost of debt (r_d)?

2. The current market price of a $1,000 par bond is $1,025. Its semiannual coupon rate is 7 percent, and it matures in five years. What is the pre-tax cost of debt (r_d)?

3. What is the after-tax cost of debt using the data in Problem 1 and assuming a 40 percent marginal tax rate?

4. What is the after-tax cost of debt using the data in Problem 2 and assuming a 40 percent tax rate?

5. What is the cost of preferred stock financing if it pays a constant dividend of $3.00 and is currently selling for $35.00 per share?

6. What is the after-tax cost of preferred using the data from Problem 5?

7. Use CAPM to compute the cost of equity given that the firm's beta is 1.1, the risk-free rate is 6 percent, and the return on the market is 13 percent.

8. Use the Gordon constant-growth model to compute the cost of equity given that the next dividend is expected to be $2.00, the current price is $30.00, and the expected growth rate is 8 percent.

9. The LAST dividend paid was $1.50. What is the cost of equity using the Gordon constant-growth model assuming current price of $25.00 and an expected growth rate of 7 percent?

10. Assume you have equal faith in each method used to compute the cost of equity. Use the results from Problems 7 and 8 to compute the cost of equity.

11. Again using the solutions to Problems 7 and 8, find the cost of equity assuming you are twice as confident in the CAPM approach as you are in the constant-growth approach to finding the cost of equity.

12. A firm currently has long-term debt totaling $1,500,000, preferred stock totaling $250,000 and equity outstanding of $3,000,000. Using the answers to Problems 4, 5, and 11, compute the weighted average cost of capital after tax.

Answers to Quick Drill Self Test

1. 8.89%

2. 6.41%

3. 5.33%

4. 3.85%
5. 8.57%
6. 8.57%
7. 13.70%
8. 14.67%
9. 13.42%
10. 14.19%
11. 14.02%
12. 10.52%

■ Study Tips

1. Recognize that there is only one generally accepted method to find the cost of debt. This is because you can be confident of the inputs to this model. There is also only one method for finding the cost of preferred stock. There are multiple methods for finding the cost of common equity because of the uncertainty of the inputs. Be sure to know the alternative methods.

2. When computing the cost of debt, be sure to multiply by (1 – tax rate) to get the after-tax cost. The after-tax cost will be given in some problems. Be alert to the wording of the problem.

3. There are two steps to the calculation of the weighted average cost of capital. First, compute the cost of each component, and then find the weighted average.

■ Student Notes

■ Sample Exam—Chapter 9

True/False

T F 1. The cost of capital is the rate of return a firm must earn on investments in order to leave share price unchanged.

T F 2. If risk is unchanged, the undertaking of projects with a rate of return above the cost of capital will decrease the value of the firm.

T F 3. The specific cost of each source of financing is viewed on a before-tax basis.

T F 4. The net proceeds used in calculating the cost of long-term debt are funds actually received from the sale after paying for flotation costs.

T F 5. When the net proceeds from the sale of a bond equal its par value, the coupon interest rate will be the bond's before-tax cost of capital.

T F 6. The cost of preferred stock is typically lower than the cost of long-term debt because dividends paid on preferred stock are tax deductible.

T F 7. The cost of common stock equity may be measured using either the zero-growth valuation model or the CAPM.

T F 8. Due to the absence of flotation costs, the cost of retained earnings is always lower than the cost of a new issue of common stock.

T F 9. The CAPM describes the relationship between the required return and the nonsystematic risk of the firm as measured by the beta coefficient.

T F 10. Larger volumes of new financing are associated with greater risk and lead to higher financing costs.

T F 11. Because preferred stock is a form of ownership, the stock will never mature.

T F 12. When calculating the WACC, the preferred weighting scheme in the target market value proportions.

T F 13. When making a capital budgeting decision, the overall cost of capital rather than just the cost of any single source of financing must be used to capture all of the relevant financing costs.

Multiple Choice

1. The _________ is the rate of return a firm must earn on its investment in order to maintain the market value of its stock.
   a. gross profit margin
   b. internal rate of return
   c. net present value
   d. cost of capital

2. The WACC represents the average _________ for the firm.
   a. cost of equity
   b. return on fixed assets
   c. cost of financing
   d. return on investment

3. The cost of capital reflects the cost of funds
   a. over the short run.
   b. at current book values.
   c. set by the federal government.
   d. over the long run.

4. The firm's optimal mix of debt and equity is called its
   a. target capital structure.
   b. maximum wealth ratio.
   c. optimal mix.
   d. debt-to-equity ratio.

5. The specific cost of each source of long-term financing is based on _________ and _________ costs.
   a. before-tax; current
   b. after-tax; historical
   c. after-tax; current
   d. before-tax; historical

6. A tax adjustment must be made in determining the cost of
   a. common stock.
   b. long-term debt.
   c. retained earnings.
   d. preferred stock.

7. A firm has issued 8 percent preferred stock, which sold for $100 per share par value. The flotation costs of the stock equaled $3, and the firm's marginal tax rate is 40 percent. The cost of the preferred stock is
   a. 8.25 percent.
   b. 7.50 percent.
   c. 7.35 percent.
   d. 9.85 percent.

8. The approximate before-tax cost of debt for a 20-year, 9 percent, $1,000 par value bond selling at $950 is
   a. 10.63 percent.
   c. 7.45 percent.
   b. 11.39 percent.
   d. 9.49 percent.

9. The cost of common stock equity may be estimated by using the
   a. IRR.
   b. NPV.
   c. constant-growth valuation model.
   d. DuPont model.

10. The cost of retained earnings is equal to
    a. the cost of long-term debt.
    b. the cost of common stock equity.
    c. zero.
    d. the marginal cost of capital.

11. A firm has a beta of 0.90. The market return equals 12 percent, and the risk-free rate of return equals 4 percent. The estimated cost of common stock equity is
    a. 11.2 percent.
    b. 9.8 percent.
    c. 10.4 percent.
    d. 12.6 percent.

12. One major expense associated with issuing new shares of common stock is
    a. legal fees.
    b. selling fees.
    c. registration fees.
    d. overpricing.

13. A firm has common stock with a market price of $45 per share and an expected dividend of $3 per share at the end of the coming year. The growth rate in dividends has been 4 percent. The cost of the firm's common stock equity is
    a. 9.75 percent.
    b. 10.67 percent.
    c. 8.42 percent.
    d. 11.25 percent.

14. Generally the least expensive form of long-term capital is
    a. short-term debt.
    b. retained earnings.
    c. long-term debt.
    d. common stock.

15. A firm has determined its cost of each source of capital and optimal capital structure, which is composed of the following sources:

Source of Capital	Proportion	After-Tax Cost
Long-term debt	45%	7%
Preferred stock	15%	10%
Common stock equity	40%	14%

The weighted average cost of capital is

a. 0.25 percent.
b. 11.45 percent.
c. 9.75 percent.
d. 8.35 percent.

16. A firm's before-tax cost of long-term debt 10.45 percent. What is the firm's after-tax cost of long-term debt if the firm has a 40 percent corporate tax rate?

a. 8.48 percent
b. 6.27 percent
c. 5.32 percent
d. 9.75 percent

17. For which source of funds does the firm not incur flotation costs?

a. common stock
b. preferred stock
c. long-term debt
d. retained earnings

18. A firm has discovered that its retained earnings of $400,000 will soon be exhausted. What is the point at which the firm will no longer be able to sustain the retained earnings cost of 6 percent if they have a historical weight of 40 percent in the firm's WACC?

a. $750,000
b. $160,000
c. $1,000,000
d. $100,000

19. When determining the after-tax cost of a bond, the face value of the bond must be adjusted to the net proceeds amounts by considering

a. risk.
b. flotation cost.
c. taxes.
d. returns.

20. When the face value of a bond equals its selling price, the firm's cost of the bond will be equal to

a. the coupon interest rate.
b. the firm's WACC.
c. the risk-free rate.
d. the firm's WMCC.

21. An individual's average cost of borrowing can be referred to as their:
    a. internal rate of return.
    b. payback period.
    c. personal cost of capital.
    d. risk-adjusted required rate of return.

22. In the process of buying a motorcycle, Joshua Brent is comparing the interest rate offered by the motorcycle shop with a second mortgage on his home. The motorcycle shop is offering 7.0 percent, while the interest rate on his second mortgage would be 8.0 percent. Given that Joshua is in the 25 percent tax bracket, which financing source should be used?
    a. the motorcycle shop, because the interest rate is 1.0 percent lower
    b. the motorcycle shop, which would result in a cost of 5.25 percent after-taxes
    c. the motorcycle shop, which would result in a cost of 8.75 percent after-taxes
    d. a second mortgage on his house, which would have a cost of 6.0 percent after-taxes

23. Philip Munson currently has two outstanding loans. One is an 8.2 percent loan with an outstanding balance of $15,000. The other is a 9.6 percent loan with an outstanding balance of $35,000. Philip's banker has allowed him to combine these loans into one with an interest rate of 9 percent. What should Philip do?
    a. Keep the current loans, which have a weighted average cost of 8.62 percent.
    b. Keep the current loans, which have a weighted average cost of 8.90 percent.
    c. Keep the current loans, which have a weighted average cost of 9.18 percent.
    d. Accept the bank's offer of money at 9.0 percent.

24. A firm has determined its cost of each source of capital and optimal capital structure, which is composed of the following sources:

Source of Capital	**Market Value**	**After-Tax Cost**
Long-term debt	20,000,000	7%
Preferred stock	5,000,000	10%
Common stock equity	25,000,000	14%

    The weighted average cost of capital is
    a. 0.11 percent.
    b. 10.80 percent.
    c. 5.50 percent.
    d. 8.35 percent.

25. A firm has common stock with a market price of $20 per share and an expected dividend of $1 per share at the end of the coming year. The growth rate in dividends has been 4 percent. Floatation costs are 5 percent. The cost of the firm's new common stock equity is
    a. 9.00 percent.
    b. 10.67 percent.
    c. 9.26 percent.
    d. 11.25 percent.

Essay

1. Why should a firm use an average weighted cost of capital rather than just the cost of any single source of financing?

2. Explain why the cost of retained earnings is lower than the cost of external equity.

■ Chapter 9 Answer Sheet

True/False

1. T
2. F
3. F
4. T
5. T
6. F
7. F
8. T
9. F
10. T
11. T
12. T
13. T

Multiple Choice

1. D
2. C
3. D
4. A
5. C
6. B
7. A $\dfrac{\$8}{(\$100-\$3)} = 8.25\%$
8. D $\dfrac{\$90 + \dfrac{\$1{,}000 - \$950}{20}}{\dfrac{\$950 + \$1{,}000}{2}} = 9.49\%$
9. C
10. B
11. A $0.04 + [0.90 \times (0.12 - 0.04)] = 11.20\%$
12. B
13. B $\dfrac{\$3}{\$45} + 0.04 = 10.67\%$
14. C
15. A $(0.45 \times 0.07) + (0.15 \times 0.10) + (0.40 \times 0.14) = 10.25\%$
16. B $10.45 \times 0.60 = 6.27\%$
17. D
18. C $\dfrac{\$400{,}000}{0.40} = \$1{,}000{,}000$
19. B
20. A
21. C
22. D
23. D
24. B $(0.4 \times 7\%) + (0.1 \times 10\%) + (0.5 \times 14\%) = 10.8\%$
25. C $r_s = (1/19) + 0.04$

Essay

1. It a firm only considered a single cost of financing, it may cause them to incorrectly accept or reject future projects. For example, consider a firm with 50 percent debt with an after-tax cost of 5 percent and 50 percent equity with a cost of 10 percent. If we were considering a project that would return 8 percent and only considered equity financing, we would reject the project but would expect it with only debt or with the weighted average cost of 7.5% (= 0.5 × 5% + 0.5 × 10%).

2. External equity is more costly because the firm will incur flotation costs in selling stock in the market. These costs include selling fees and underpricing.

Chapter 10
Capital Budgeting Techniques

■ Chapter Summary

In this chapter, the tools you have been developing since Chapter 5 come together to help evaluate projects. The main methods for evaluating projects are the net present value and the internal rate of return. You already know the mechanics necessary to compute both of these numbers. This chapter discusses the theory, advantages, and disadvantages of each method. The next chapter provides techniques for dealing with some of the problems that arise in their application. While you may not find this chapter as difficult as others, it is among the most valuable.

LG 1 *Understand the key elements of the capital budgeting process.* Capital budgeting is the process of evaluating and selecting investment opportunities. Capital budgeting techniques provide a method for using TVM to help select among competing investments.

LG 2 *Calculate, interpret, and evaluate the payback period.* The payback period is the amount of time required for the firm to recover its initial investment, as calculated from cash inflows. Shorter payback periods are preferred. The payback period is relatively easy to calculate, has simple intuitive appeal, considers cash flows, and measures risk exposure. Its weaknesses include lack of linkage to the wealth maximization goal, failure to consider time value explicitly, and the fact that it ignores cash flows that occur after the payback period.

LG 3 *Calculate, interpret, and evaluate the net present value (NPV) and economic value added (EVA).* You will not find it difficult to learn to use *NPV* to evaluate projects because this method is just an extension of the time value of money principles learned in Chapter 5. Because it gives explicit consideration to the time value of money, NPV is considered a sophisticated capital budgeting technique. NPV measures the amount of value created by a given project; only positive NPV projects are acceptable. The rate at which cash flows are discounted in calculating NPV is called the discount rate, required return, cost of capital, or opportunity cost. By whatever name, this rate represents the minimum return that must be earned on a project to leave the firm's market value unchanged. Be sure to use time lines for evaluating projects with more complex cash flows.

LG 4 *Calculate, interpret, and evaluate the internal rate of return* (*IRR*). Like NPV, IRR is a sophisticated capital budgeting technique. IRR is the compound annual rate of return the firm will earn by investing in a project and receiving the given cash inflows. By accepting only those projects with IRRs in excess of the firm's cost of capital, the firm should enhance its market value and the wealth of its owners. Both NPV and IRR yield the same accept–reject decisions, but they often provide conflicting ranking. The IRR is computed in the same way you learned to compute the yield to maturity for bonds. The number of IRRs equals the number of sign changes in its cash flows.

LG 5

Use net present value profiles to compare NPV and IRR techniques. A net present value profile is a graph that shows the net present value at a variety of different costs of capital. The value of the NPV profile is that it shows the sensitivity of NPV to the selection of the discount rate. The IRR can be read directly off the NPV profile where the graph crosses the *x*-axis, depicting NPV equal to zero.

LG 6

Discuss NPV and IRR in terms of conflicting rankings and the theoretical and practical strengths of each approach. NPV and IRR will always provide the same accept/reject decision. However, they will not always rank projects the same as a result of differences in reinvestment rate assumption, as well as the magnitude and timing of cash flows. This conflict becomes important when a firm cannot afford to accept both projects or when it is trying to select among mutually exclusive projects.

■ Chapter Notes

Capital Budgeting Techniques

LG 1

Capital budgeting techniques are used by firms to evaluate projects that will increase shareholders' wealth. The preferred techniques combine time value procedures, risk and return considerations, and valuation concepts to select the projects that are most consistent with the firm's goals. The three most popular capital budgeting techniques are the payback period, net present value, and internal rate of return.

Capital Expenditure Motive

A capital expenditure is an outlay of funds by the firm that is expected to produce benefits over a period of time greater than one year. The motives for capital expenditures vary; the following table describes the key motives for making capital expenditures.

Motive	Description
Expansion	The most common motive for capital expenditure. By expanding the level of operations for a growing firm, the firm finds it necessary to acquire fixed assets.
Replacement	As a firm's growth reaches maturity, a majority of its capital expenditures will be for the replacement of worn-out assets.
Renewal	Instead of replacing an obsolete asset, sometimes it may be more cost effective to overhaul or repair the old asset.
Other Purposes	Some capital expenditures do not clearly fall into one of the previous categories. Often firms outlay capital for intangible investments such as advertising, research and development, and management consulting.

Steps in the Process

The capital budgeting process is made up of five steps. The process begins with proposal generation followed by review and analysis, decision making, implementation, and follow-up. The steps in the capital budgeting process are summarized in the following table.

Steps	Description
Proposal Generation	Proposals for capital expenditures are made at all levels within an organization. Proposals often travel from an originator at a lower level in the organization to a reviewer at a higher level in the organization.
Review and Analysis	Capital expenditure proposals are reviewed on the basis of two criteria: (1) to assess their appropriateness in light of the firm's overall objectives and plans and (2) to evaluate their economic validity. When the analysis is completed, a summary and recommendation is submitted to the decision maker.
Decision Making	The final decision on proposed capital expenditures is often made on the basis of dollar amount. Clearly, the most expensive capital expenditures will be left to the board of directors, while inexpensive items, such as a $5 trash can, will be handled by the lowest levels of management.
Implementation	Once a proposal has been approved and funds have been made available, the implementation phase begins. The implementation of minor outlays often begins immediately, while the implementation of major outlays may occur in phases.
Follow-up	Follow-up involves monitoring the results of the capital expenditure during the operating phase of the project. When actual costs deviate from projected costs, action may be needed to improve efficiency and cut costs.

Basic Terminology

Independent *projects* are projects where cash flows are unrelated or independent from one another. The acceptance of one project does not necessarily eliminate other projects.

Mutually exclusive projects are projects that are competing with one another. The acceptance of a mutually exclusive project automatically eliminates other mutually exclusive projects.

When a firm has unlimited funds, all independent projects that provide returns greater than the firm's hurdle rate will be accepted. However, most firms are not in this situation. Most firms use *capital rationing*, which means that the firm must ration its funds by allocating them to projects that will maximize share value. Under capital rationing, numerous projects will compete for limited dollars.

There are two basic approaches to capital budgeting. The *accept–reject approach* allows firms to either accept or reject projects that are brought before them based on some minimum acceptance criteria. The *ranking approach* ranks projects on the basis of some predetermined measure such as the rate of return.

Conventional cash flow patterns consist of an initial outflow of cash in year 0 followed by a series of inflows. A nonconventional *cash flow* pattern is one in which an initial outflow is not followed by a series of inflows. For example, a machine is purchased in year 0 for $10,000. The machine provides $5,000 of cash inflows in years 1, 2, 3, and 4. In year 3 the machine must be overhauled at a cost of $6,000. The outflow of $6,000 in year 3 breaks the pattern of steady inflows.

An *annuity* is a stream of equal cash flows. A series of cash flows that exhibit any pattern different from an annuity is termed a *mixed stream* of cash flows.

Payback Period

LG 2

The payback period is the exact amount of time required for a firm to recover its initial investment in a project as calculated from cash inflows. The payback period is popular because of its ease in calculation, but it is not often used because the appropriate payback period is subjective, it fails to recognize the time value of money, and it does not recognize cash flows after the payback period.

Net Present Value

LG 3

Net present value is the preferred technique because it gives consideration to the time value of money. The net present value is found by subtracting the initial investment from the present value of the cash inflows discounted at a rate equal to the firm's cost of capital. The formula for NPV is shown in the following equation.

$$NPV = \sum_{t=1}^{n} \frac{CF_t}{(1+r)^t} - CF_0$$

where:

CF_t = individual year's cash flow

r = firm's cost of capital

CF_0 = initial investment

If the NPV of a project is greater than \$0, the project should be accepted, and if the NPV is less than \$0, the project should be rejected.

NPV and the Profitability Index

A variation of the NPV rule.

$$PI = \frac{\sum_{t=1}^{n} \frac{CF_t}{(1+r)^t}}{CF_0}$$

When companies evaluate investment opportunities using the PI, the decision rule they follow is to invest in the project when the index is greater than 1.0.

NPV and Economic Value Added

EVA calculates a "cost of capital" charge, which is deducted each year from a project's cash flows. To calculate the overall project EVA, you take the annual EVA figures and discount them at the cost of capital. In general, NPV, PI, and EVA will always agree on whether a project is worth investing in or not.

Internal Rate of Return (IRR)

LG 4

The internal rate of return (IRR) is probably the most widely used capital budgeting technique. The IRR is defined as the discount rate that equates the present value of cash inflows with the initial investment associated with a project. If the IRR of a project is greater than the firm's cost of capital, the project should be accepted, and if the IRR is less than the cost of capital, the projects should be rejected. The equation used to find a project's IRR is shown in the following formula.

$$\sum_{t=1}^{n} \frac{CF_t}{(1+r)^t} = CF_0$$

Comparing NPV and IRR Techniques

LG 5

The NPV and IRR often rank projects differently. The NPV assumes the reinvestment of funds at the firm's cost of capital, while the IRR assumes the reinvestment of funds at a rate equal to the project's IRR.

The net present value profile can help you to understand why NPV and IRR can rank projects differently. In the following figure, the NPV of Projects A and B have been computed at a variety of discount rates. At a 5 percent discount rate, Project A is preferred because it has the higher NPV. At a 10 percent discount rate, Project B is preferred because it has the higher NPV. This example shows that the preference depends on the discount rate. If IRR were used to rank projects, it would always rank B as best. This is because the project's IRR is where it crosses the *x* axis, and B crosses further out.

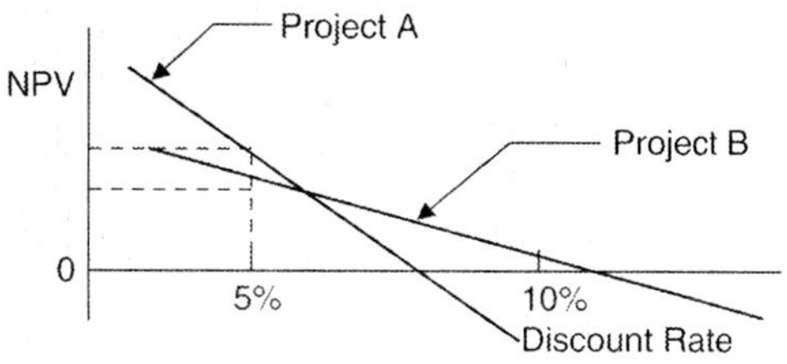

LG 6

Theoretically, the NPV is the better approach to capital budgeting. The NPV's assumption of reinvestment at the firm's cost of capital is more realistic than the IRR's assumption of reinvestment at the project's IRR. But there is evidence that suggests some financial managers prefer the IRR technique. Most business people are more concerned with rates of return than actual dollar returns.

Sample Problems and Solutions

Example 1. Payback

Suppose that you want to invest in a $100,000 project that you expect to provide cash inflows of $50,000 in the first year, $25,000 the second, and $20,000 per year thereafter. What is the payback period?

Solution

The payback period is simply the number of years that it takes to recover your investment. Note that payback ignores the timing and riskiness of the cash flows, as well as any cash flows that are received after the payback period has passed. Compute payback by determining the cumulative cash flows received by year.

Year	Cash Flow	Cumulative Cash Flow
Year 1	$50,000	$50,000
Year 2	25,000	75,000
Year 3	20,000	95,000
Year 4	20,000	(1/4) 5,000

After 3.25 years, the investment of $100,000 will be received. Thus the payback is 3.25 years.

Example 2. Net Present Value

How would you explain what an NPV of $10,000 meant to someone who had not taken any finance?

Solution

You would explain that the $10,000 was how much more was earned by this project, in today's dollars, than was necessary to compensate you for the risk of the project.

Example 3.

If your company was not subject to capital rationing, would you accept a project that had a net present value of $1.00, even if the initial investment was $1 billion?

Solution

Yes. Refer to the answer to Example 2. An NPV of $1.00 says that you have received everything you needed to be adequately compensated for the risk of the project, plus $1.00 more. The problem does not say what the rate of return may have been. If the discount rate was 25 percent, then you will have earned 25 percent compounded every year, plus an extra dollar.

Example 4.

You estimate the net initial investment of a project to be $250,000. Annual cash inflows of $50,000 per year for eight years are expected to follow. The firm's cost of capital is 12 percent. Should you invest?

Calculator Solution

Refer to the financial calculator section in the back of the study guide for detailed keystroke examples.

$$CF_0 = -250{,}000$$
$$C_{01} = 50{,}000$$
$$F_{01} = 8$$
$$I = 12$$
$$\text{NPV} = -\$1{,}618.01$$

Because the net present value is less than 0, you should not invest in this project.

Example 5.

You estimate the relevant net cash inflows on an investment to be $20,000 per year for six years. The initial investment is $60,000 and the cost of capital is 10 percent. Should you invest?

Calculator Solution

$$CF_0 = -60{,}000$$
$$C_{01} = 20{,}000$$
$$F_{01} = 6$$
$$I = 10$$
$$\text{NPV} = \$27{,}105.21$$

Because the NPV is positive, you should invest.

Example 6.

Assume that you are evaluating an investment that requires you to invest $10,000 per year for three years. Following this you will not receive anything for two years, then you will receive $20,000 for the next three years. If the cost of capital is 12 percent, should this project be accepted?

a. Table Solution

If the initial cash flow is disbursed over a period spanning several years, find the present value of the cash outflows and subtract this amount from the present value of the cash inflows.

It is often best to draw a time line when the cash flows become complicated.

−10,000	−10,000	−10,000	0	0	20,000	20,000	20,000
0	1	2	3	4	5	6	7

b. Calculator Solution

$$CF_0 = -10{,}000$$
$$C_{01} = -10{,}000$$
$$F_{01} = 2$$
$$C_{02} = 0$$
$$F_{02} = 2$$
$$C_{03} = 20{,}000$$
$$F_{03} = 3$$
$$I = 12$$
$$\text{NPV} = \$3{,}627.63$$

Because the NPV is positive, the investment should be accepted.

Example 7.

You are asked to evaluate two projects, X and Y. The cash flow for each is listed below. The firm's cost of capital is 12 percent. Using the NPV method, which project would you select? Why?

Year	X	Y
0	(10,000)	(10,000)
1	6,500	3,500
2	3,000	3,500
3	3,000	3,500
4	1,000	3,500

a. Formula Solution

$$\text{NPV}_x = \$6{,}500/1.12 + \$3{,}000/1.12^2 + \$3{,}000/1.12^3 + \$1{,}000/1.12^4 - \$10{,}000 = \$966$$
$$\text{NPV}_y = \$3{,}500/1.12 + \$3{,}500/1.12^2 + \$3{,}500/1.12^3 + \$3{,}500/1.12^4 - \$10{,}000 = \$630$$

Because the NPV of project X is greater than the NPV of Project Y, accept X.

b. Calculator Solution

X	Y
$CF_0 = -10{,}000$	$CF_0 = -10{,}000$
$C_{01} = 6{,}500$	$C_{01} = 3{,}500$
$F_{01} = 1$	$F_{01} = 4$
$C_{02} = 3{,}000$	$I = 12$
$F_{02} = 2$	$\text{NPV}_y = \$630.72$
$C_{03} = 1{,}000$	
$F_{03} = 1$	
$I = 12$	
$\text{NPV}_x = \$966.01$	

Example 8. Internal Rate of Return

Compute the internal rate of return for Project Y in Example 7 above.

Solution

The internal rate of return is the interest rate that sets the present value of the cash inflows equal to the present value of the cash outflows.

a. Calculator Solution

$$PV = -10{,}000$$
$$FV = 0$$
$$\text{PMT} = 3{,}500$$
$$N = 4$$
$$\text{CPT } I = 14.96\%$$

Therefore, the IRR of Project Y is 14.96 percent. Because this is greater than the cost of capital of 12 percent, the project should be accepted. Note that the decision reached using the IRR method is the same as what was reached in Example 7 using NPV. This will always be the case. If computed properly, IRR and NPV will always give the same accept/reject decision.

Example 9. Net Present Value Profile

Assume that you want to evaluate an investment that requires you to pay out \$15,000 now and that promises to return \$4,000 each year for five years. Compute the net present value profile for this project.

Solution

To compute a NPV profile you need to find the NPV at a variety of discounts rates, then plot the result. It is not too important which discount rates you choose to use. Since the calculation of NPV is very easy for 0 percent discount, be sure to begin there. You then want to choose discount rates that result in some positive and some negative NPVs.

Discount Rate	NPV
0.0%	\$5,000
2.5%	\$3,583
5.0%	\$2,318
7.5%	\$1,184
10.0%	\$ 163
12.5%	–\$ 758
15.0%	–\$1,591
17.5%	–\$2,348

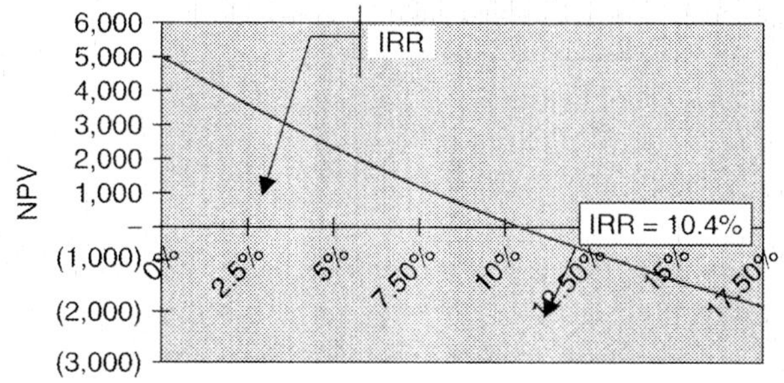

Once the NPVs and discount rates are plotted, it is easy to find the IRR. This is where the NPV is equal to zero. The steeper the NPV profile, the more sensitive the investment is to the discount rate assumption.

Example 10.

You are asked to evaluate two projects, Q and P. The cash flow for each is listed below. The firm's cost of capital is 15 percent. Calculate the NPV, IRR, and PI for both projects and then rank the projects based on their NPV, IRR, and PI. Do the rankings agree? If not, why not? Which project would you recommend and why?

Year	P	Q
0	(30,000)	(30,000)
1	12,250	11,500
2	13,000	11,500
3	10,000	11,500
4	9,500	11,500

a. Formula Solution

$NPV_P = \$12{,}250/1.15 + \$13{,}000/1.15^2 + \$10{,}000/1.15^3 + \$9{,}500/1.15^4 - \$30{,}000 = \$2{,}706.25$

$NPV_Q = \$11{,}500/1.15 + \$11{,}500/1.15^2 + 11{,}500/1.15^3 + \$11{,}500/1.15^4 - \$30{,}000 = \$2{,}832.25$

Because the NPV of project Q is greater than the NPV of Project P, choose Q.

b. Calculator Solution

P	**Q**
$CF_0 = -30{,}000$	$CF_0 = -30{,}000$
$C_{01} = 12{,}250$	$C_{01} = 11{,}500$
$F_{01} = 1$	$F_{01} = 4$
$C_{02} = 13{,}000$	$I = 15$
$F_{02} = 1$	$NPV_Q = 2{,}832.25$
$C_{03} = 10{,}000$	
$F_{03} = 1$	
$C_{04} = 9{,}500$	
$F_{04} = 1$	
$I = 15$	
$NPV_P = 2{,}706.25$	

P	**Q**
$CF_0 = -30{,}000$	$CF_0 = -30{,}000$
$C_{01} = 12{,}250$	$C_{01} = 11{,}500$
$F_{01} = 1$	$F_{01} = 4$
$C_{02} = 13{,}000$	$IRR_Q = 19.60\%$
$F_{02} = 1$	
$C_{03} = 10{,}000$	
$F_{03} = 1$	
$C_{04} = 9{,}500$	
$F_{04} = 1$	
$IRR_P = 19.65\%$	

Because the IRR of project P is greater than the IRR of Project Q, choose P.

Compute PI: $PI_P = 32{,}706.25/30{,}000 = 1.0902$; $PI_Q = 32{,}832.25/30{,}000 = 1.0944$

Because the PI of project Q is greater than the NPV of Project P, choose Q.

The rankings do not agree. Using NPV and PI project Q should be chosen; however, using IRR project P should be chosen. This is because the IRR method assumes reinvestment at the IRR rate, note the required return. The higher NPV project, project Q, should be chosen because the shareholders would be better off.

■ Quick Drill Self Test

1. XYZ Company wants to know the payback period for a project with an initial investment of $4 million and annual cash flows of $800,000.

2. Suppose the annual cash flows listed for Question 1 start at $800,000 and then decrease by 15 percent each year. What is the payback period?

3. A firm is evaluating a project with an initial cost of $3.35 million and annual cash flows of $1.15 million for 4 years. If the cost of capital for the firm is 14 percent, what is the NPV? Should the firm accept or reject the project?

4. Suppose the cost of capital for the firm in Question 3 increases to 15 percent, what is the NPV? Should the firm accept or reject the project?

5. Evergreen Inc. is evaluating a project with an initial cost of $6 million. Cash flows would start at $1 million and increase by $750,000 annually for the next 3 years. If the cost of the capital for the firm is 12 percent, what is the NPV? Should the firm accept or reject the project?

6. A project has projected cash outflows of $2 million in the current time period. Additional cash outflows of $1 million, $1 million, and $2 million are projected during the next three years of operation. Cash inflows of $1.3 million are expected in years 4 through 13 (10 cash inflows). Should the project be accepted if the company has a cost of capital of 14 percent? (What is the NPV?)

7. Would the accept–reject decision change for the project described in Question 6 if the cost of capital fell to 12 percent?

8. Suppose that a $10,000 investment will yield 3 cash flows of $4,000 each. With a discount rate of 12 percent, what is the PI? What is the accept–reject decision?

9. Rank the following projects by PI.

Project	**Net Investment**	***PV* (Cash Inflows)**	**NPV**
A	$400	$480	$80
B	700	735	35
C	150	225	75
D	1,000	950	(50)
E	250	275	25

10. If the initial investment for a project is $500 and the cash inflows are $300 for three years, what is the IRR?

11. If the initial investment for a project is $500 and the cash inflows are $300 for year 1, $250 for year 2, and $150 for year 3, what is the IRR?

12. If the initial investment for a project is $1,500 and the cash inflows are $300 for four years, what is the IRR?

13. Suppose Project A has a NPV of $350 with an IRR of 12 percent and Project B has a NPV of $300 with an IRR of 20 percent. If these projects are mutually exclusive, which project should you accept? Why?

14. What is the NPV of a project with initial investments of $3 million at the beginning of years 1 and 2, and cash inflows of $1.5 million at the beginning of years 2 through 6? Assume the cost of capital is 10. Should the project be accepted? (*Hint:* At the beginning of year 2 the net cash flow is −1.5 million.)

Answers to Quick Drill Self Test

1. 5 years
2. 9.6 years
3. $769; accept
4. –$66,775; reject
5. $132,831; accept
6. –$419,655; reject
7. $114,621; accept
8. 0.96; reject
9. C, A, E, B, D
10. 36.31 percent
11. 21.48 percent
12. –8.36 percent
13. Project A; the reinvestment rate assumption is a problem when you are attempting to rank mutually exclusive projects with IRR. Thus, NPV is more useful when attempting to select among mutually exclusive projects.
14. –$41,093; reject

Long Problem

Using the cash flows for an investment project given below, compute NPV and IRR.

Year	0	1	2	3	4	5
Net Cash Flow	(1,800,000)	189,400	318,500	307,600	314,390	1,705,899

1. Compute the NPV and IRR, assuming a cost of capital of 14 percent. Is the project acceptable?
2. Prepare an NPV profile. Locate the IRR and the 14 percent cost of capital on this graph. (To locate points on the graph in Excel go to View, Toolbars, and click on Drawing. Select Autoshapes.)
3. Prepare a sensitivity profile looking at different assumed initial sales amounts. Use some estimates above the original estimate and some below.
4. What final sales price makes this project just acceptable using the assumptions provided used in Question 2? (In Excel go to Tools, Goal Seek, and input that you want the value of NPV to be zero by changing the final purchase price.)

Solution Prepared Using Excel

NPV	(109,028)				
IRR	0.1216				
PI	0.9394				
IRR	0.1216				
Discount Rate	0.14				
Growth Rate		0.10	0.07	0.05	0.05

Cost of Capital	NPV	Sales	NPV
0.00	1,035,789	450,000	(223,728)
0.03	716,453	500,000	(109,028)
0.05	530,252	550,000	5,671
0.08	284,708	600,000	120,371
0.10	140,471		
0.12	(5,741)		
0.14	(109,028)		

Sensitivity to the initial sales estimate is found by plugging in different values for sales and recording the resulting NPV. The project's sensitivity to the cost of capital and to sales is graphed below.

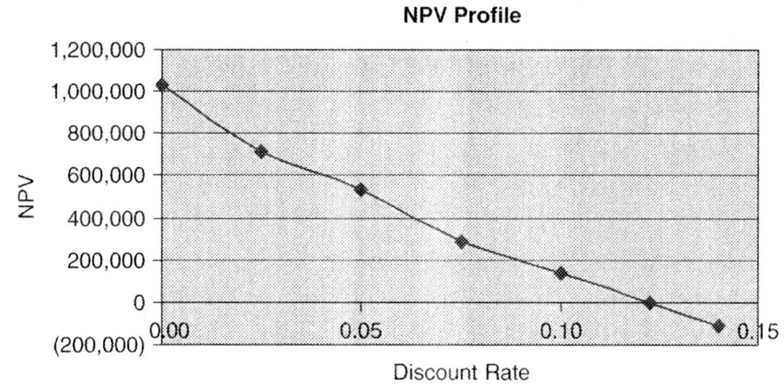

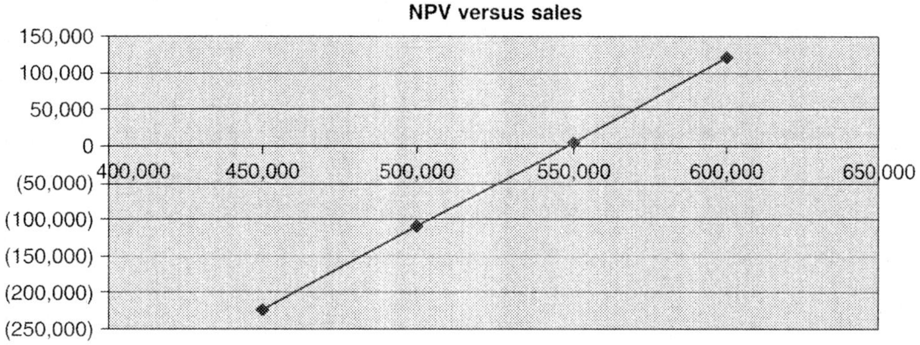

■ Study Tips

1. Learn the advantages and disadvantages of the NPV and IRR approaches. This question appears on virtually every exam given on this chapter.

2. Whatever the cashflow stream may be, the NPV will simply be the present value. Be careful to keep track of all of the cashflow signs.

3. Whenever analyzing complicated cash flows, draw a time line so that you will correctly count the periods back to time 0.

4. Remember not to use the IRR to rank projects. The reinvestment interest rate assumption may cause you to rank projects incorrectly. If you need to rank projects for firms experiencing capital rationing, use NPV.

■ Student Notes

____

____

____

____

____

■ Sample Exam—Chapter 10

True/False

T F 1. The payback period is the amount of time, rounded to the nearest year, that is required for a firm to recover the cost of a new asset.

T F 2. Net present value is considered a sophisticated capital budgeting technique because it gives consideration to the time value of money.

T F 3. The internal rate of return is the discount rate that equates the present value of the cash inflows of a project with its initial investment.

T F 4. If the NPV of a project is zero, the IRR of that project will always be less than the firm's cost of capital.

T F 5. The goal of the firm should be to use its budget to generate the highest possible internal rate of return for its cash inflows.

T F 6. The NPV assumes that intermediate cash inflows are invested at a rate equal to the firm's cost of capital.

T F 7. The NPV profile shows how sensitive a project's NPV is to changes in the discount rate.

T F 8. The IRR can be read directly off an NPV profile as the point where the profile crosses the x-axis.

T F 9. For conventional projects, the NPV and the IRR will always produce the same accept–reject decision.

T F 10. The internal rate of return assumes that the intermediate cash flows associated with a project will be reinvested at the project's IRR.

T F 11. Mutually exclusive projects have the same function and compete with one another.

T F 12. Under capital rationing, firms have an unlimited amount of funds for investments.

T F 13. Independent projects are those whose cash flows are unrelated to one another.

Multiple Choice

1. The first step in the capital budgeting process is
   a. proposal generation.
   b. analysis.
   c. review.
   d. decision making.

2. A project having the conventional pattern of cash flows exhibits all of the following EXCEPT
   a. operating cash inflows.
   b. a terminal cash flow.
   c. operating cash outflows.
   d. initial investment.

3. A conventional cash flow pattern associated with capital investment projects consists of an initial outflow followed by a
   a. broken series of outflows.
   b. broken series of inflows.
   c. series of outflows.
   d. series of inflows.

4. The final step in the capital budgeting process is
   a. implementation.
   b. follow-up.
   c. review.
   d. analysis.

5. __________ projects do not compete with one another, so the acceptance of one will have no bearing on the other projects being considered by the corporation.
   a. Independent
   b. Mutually exclusive
   c. Capital
   d. Replacement

6. Individuals can approach the acquisition process much as corporations do, using a five-step process. In which step is the purchase made?
   a. Step 1
   b. Step 3
   c. Step 4
   d. Step 5

7. The _________ is the exact amount of time it takes the firm to recover its initial investment.
   a. internal rate of return
   b. net present value
   c. payback period
   d. certainty equivalent

8. A firm is evaluating a proposal that has an initial investment of $45,000 and has cash flows of $5,000 in year 1, $20,000 in year 2, $15,000 in year 3, and $10,000 in year 4. The payback period of the project is
   a. 3.5 years.
   b. 3 years.
   c. 4 years.
   d. 2.5 years.

9. All of the following are examples of sophisticated capital budgeting techniques EXCEPT
   a. net present value.
   b. annualized net present value.
   c. internal rate of return.
   d. payback period.

10. The _________ is the discount rate that equates the present value of the cash inflows with the initial investment.
    a. cost of capital
    b. internal rate of return
    c. average rate of return
    d. opportunity cost

11. A firm with a cost of capital of 11 percent is evaluating four capital projects. The internal rates of return are as follows:

Project	IRR
1	13 percent
2	10 percent
3	11 percent
4	15 percent

    The firm should
    a. accept 4 and 1, and reject 2 and 3.
    b. accept 4, 1, and 3, and reject 2.
    c. accept 4, and reject 1,2, and 3.
    d. accept 3, and reject 1,2, and 4.

12. The __________ is the minimum amount of return that must be earned on a project in order to leave the firm's value unchanged.
    a. internal rate of return
    b. compound rate
    c. discount rate
    d. risk-free interest rate

13. A firm is evaluating an investment proposal that has an initial investment of $8,000 and discounted cash flows valued at $6,000. The net present value of the investment is
    a. $0.
    b. –$2,000.
    c. $2,000.
    d. $6,000.

14. Net present value and internal rate of return analyses
    a. always result in the same ranking of projects.
    b. always result in the same accept–reject decision.
    c. may result in differing rankings of projects.
    d. Both b and c are correct.

15. Unlike the IRR criteria, the NPV approach assumes an interest rate equal to the
    a. market interest rate.
    b. project's internal rate of return.
    c. risk free rate of return.
    d. firm's cost of capital.

For Questions 16, 17 and 18, refer to the following information:

A firm has undertaken a project with an initial investment of $100,000. The firm's cost of capital is 14 percent.

Year	**Cash Inflow**
1	$50,000
2	$65,000
3	$90,000

16. What is the NPV for the project?
    a. $50,000
    b. $32,486
    c. $54,622
    d. $76,549

17. What is the IRR for the project?
    a. 41 percent
    b. 46 percent
    c. 32 percent
    d. 22 percent

18. What is the PI for the project?
    a. 0.55
    b. 1.55
    c. 1.83
    d. 1.50

19. All of the following are weaknesses of the payback period EXCEPT
    a. a disregard for cash flows after the payback period.
    b. only an implicit consideration of the timing of cash flows.
    c. the difficulty of specifying the appropriate payback period.
    d. that it uses cash flows, not accounting profits.

20. In the context of capital budgeting, risk refers to
    a. the degree of variability of the cash inflows.
    b. the degree of variability of the initial investment.
    c. the chance that the net present value will be greater than zero.
    d. the chance that the internal rate of return will exceed the cost of capital.

21. Sophisticated capital budgeting techniques do not
    a. examine the size of the initial outlay.
    b. use net profits as a measure of return.
    c. explicitly consider the time value of money.
    d. take into account an unconventional cash flow pattern.

22. The amount by which the required discount rate exceeds the risk-free rate is called the
    a. opportunity cost.
    b. risk premium.
    c. risk equivalent.
    d. excess risk.

23. The minimum return that must be earned on a project in order to leave the firm's value unchanged is the
    a. internal rate of return.
    b. interest rate.
    c. discount rate.
    d. compound rate.

24. A firm would accept a project with a net present value of zero because the
    a. project would maintain the wealth of the firm's owners.
    b. project would enhance the wealth of the firm's owners.
    c. return on the project would be positive.
    d. return on the project would be zero.

25. Chris has been offered the chance to invest $120,000 in a partnership which is expected to return $25,000 per year. If Chris is in the 30 percent tax bracket and limits investments to those with a payback of 6 years, should Chris invest?

    a. Yes, because the payback period is 3.36 years.
    b. Yes, because the payback period is 4.80 years.
    c. No, because the payback period is 6.86 years.
    d. No, because the payback period is 16 years.

26. Jenna is considering an investment that has a price of $16,000. She expects to receive $1,000 for three years, followed by $1,400 for another four years. At the end of the seventh year, Jenna expects to sell the investment for $25,000. What is the investment's internal rate of return?

    a. 11.49 percent
    b. 12.85 percent
    c. 16.23 percent
    d. 30.00 percent

27. In the problem above, if Jenna can borrow money at a rate of 10 percent, what is the investment's net present value?

    a. $1,484
    b. $2,650
    c. $4,515
    d. $4,849

28. A project costs $1.5 million up front and will generate cash flows in perpetuity of $160,000. The firm's cost of capital is 10 percent, what is the EVA?

    a. $10,000
    b. –$10,000
    c. $100,000
    d. $16,000

Essay

1. Which of the following mutually exclusive projects should be accepted using NPV?

	Project A	**Project B**
Cost of Capital	10 percent	10 percent
Initial Investment	$50,000	$75,000
Cash Flow Year 1	$25,000	$25,000
Cash Flow Year 2	$30,000	$25,000
Cash Flow Year 3	$10,000	$45,000

2. In the above example, which project(s) should be accepted if the projects are independent of one another and the IRR approach is used as the accept–reject criterion?

Chapter 10 Answer Sheet

True/False

1. F
2. T
3. T
4. F
5. F
6. T
7. T
8. T
9. T
10. T
11. T
12. F
13. T

Multiple Choice

1. A
2. C
3. D
4. B
5. A
6. C
7. C
8. A
9. D
10. B
11. B
12. C
13. B (6,000 – 8,000)
14. D
15. D
16. C (*PV* of CF_1, CF_2, and CF_3 at 14 percent –$100,000)
17. A (must use a financial calculator)
18. B (PV of cash inflow/ initial investment)
19. D
20. A
21. B
22. B
23. C
24. A
25. C
26. B
27. B
28. A (160,000 – (1,500,000 × 0.1) = 10,000)

Essay

1. Project A has an NPV of $5,034. Project B has an NPV of $2,198. Since the projects are mutually exclusive, only one of the projects may be chosen. The project that would increase shareholder wealth the most would be Project A.

2. Project A provides an IRR of 16.34 percent. Project B provides an IRR of 11.49 percent. Since the projects are independent of one another and they both provide an IRR greater than the firm's cost of capital, both should be accepted.

Chapter 11
Capital Budgeting Cash Flows and Risk Refinements

■ Chapter Summary

An investment will involve putting out money today with the expectation that more money will be received in the future. These cash flows must be estimated as precisely as possible before they are evaluated using the evaluation tools introduced in the previous chapter. We will also learn a number of valuable techniques that adjust for project risk, perform scenario analysis, and evaluate projects under capital rationing and with different lives.

LG 1 *Discuss the relevant cash flows and the three major cash flow components.* To evaluate investment opportunities, financial managers must determine the relevant or incremental cash flow of the project. The cash flows of any project may include three basic components: (1) the initial investment, (2) operating cash inflows, and (3) terminal cash flows. All projects will have the first two components; however, some will not have a terminal cash flow.

LG 2 *Discuss expansion versus replacement decisions, sunk costs, and opportunity costs.* For replacement decisions, the relevant cash flows are the difference between the cash flows of the new asset and the old asset. Expansion decisions are viewed as replacement decisions in which all cash flows from the old asset are zero. When estimating relevant cash flows, ignore sunk costs and include opportunity costs as cash outflows.

LG 3 *Calculate the initial investment, operating cash inflows, and terminal cash flows associated with a proposed capital expenditure.* The initial cash flow includes the installed cost of the new asset, any after-tax proceeds from selling old assets, and any changes in the net working capital that result from beginning the project. There is typically a tax implication from the sale of an old asset, which depends on the relationship between its sale price and book value and on existing government tax rules. The operating cash inflows are the incremental after-tax cash inflows expected to result from a project. The income statement format involves adding depreciation back to net operating profit after taxes and gives the operating cash inflows, which are the same as operating cash flows (OCF), associated with the proposed and present projects. The relevant (incremental) cash inflows for a replacement project are the difference between the operating cash inflows of the proposed project and those of the present project. The terminal cash flow will include the proceeds of selling the equipment or assets used in the project. The taxes on the sale of the asset must be included. Any net working capital will be recovered. If it is a replacement project, then the after-tax cash flow of the old equipment at termination date must be included. Operating cash flows are not part of the terminal cash flow.

LG 4 *Understand the importance of recognizing risk in the analysis of capital budgeting projects and discuss risk and cash inflows, scenario analyses, and simulation as behavioral approaches for dealing with risk.* If a project is more or less risky than the normal project taken on by the firm, the firm cannot use the cost of capital in its evaluation. Doing so would result in the firm accepting projects it should reject or rejecting projects it should accept. Risk in capital budgeting is the degree of variability of cash flows, which for conventional capital budgeting projects stems almost entirely from net cash flows. Finding the breakeven cash inflow and estimating the probability that it will be realized make up one behavioral approach for assessing capital budgeting risk. Scenario analysis is another behavioral approach for capturing the variability of cash inflows and NPVs. Simulation is a statistically based approach that results in a probability distribution of project returns.

LG 5 *Describe the determination and use of risk-adjusted discount rates (RADRs) portfolio effects and the practical aspects of RADRs*. One way to adjust for risk is to increase or decrease the discount rate used in the capital budgeting process to compensate for the increased or decreased risk of a specific project compared to a firm's usual projects.

LG 6 *Select the best of a group of unequal-lived mutually exclusive projects using annualized net present values (ANPVs) and explain the role of real options and the objective and procedures for selecting projects under capital rationing.* Longer lived projects will tend to have higher NPVs than shorter ones because there is more time for the NPV to accumulate. This may bias the financial manager into accepting long projects when a series of short projects may be more profitable. The ANPV approach converts NPV of each unequal-lived project into an equivalent annual amount. Real options are opportunities that are embedded in capital projects and that allow managers to alter their cash flow and risk in a way that affects project acceptability (NPV). By explicitly recognizing real options, the financial manager can find a project's strategic NPV. Some of the more common types of real options are abandonment, flexibility, growth, and timing options. The strategic NPV improves the quality of the capital budgeting decision.

■ Chapter Notes

The Three Major Cash Flow Components

LG 1 As introduced in Chapter 10, capital budgeting is the process of evaluating and selecting long-term investments that are most likely to increase shareholder wealth. The most common long-term investment is the fixed asset. To evaluate these investment opportunities financial managers must determine the relevant cash flows associated with the project.

The Relevant Cash Flows

The relevant cash flows are the incremental cash outflows (investments) and inflows (returns). These cash flows of any project may include any of three basic components: (1) the initial investment, (2) operating cash inflows, and (3) terminal cash flows. All projects will have the first two components; however, some will not have a terminal cash flow.

The initial investment is the relevant cash outflow for a proposed project at time zero. The operating cash flows are the incremental after-tax cash inflows resulting from the implementation of a project during its life. The terminal cash flow is the after-tax nonoperating cash flow occurring in the final year of a project, usually attributable to the liquidation of the project.

Expansion versus Replacement Decisions, Sunk Costs, and Opportunity Costs

Expansion versus Replacement Decisions

LG 2 The calculation of cash flows received from expansion is quite different from the calculation of cash flows received from replacement. In expansion, the initial investment, operating cash inflows, and terminal cash flows are simply the after-tax cash inflows and outflows provided by the project. Under replacement, the initial investment is the cost of acquiring the new asset plus the after-tax liquidation of the asset being replaced. The operating cash flows are the difference between the cash flows provided by the new asset and the cash flows that would have been provided by the old asset. The terminal cash flow can be found by taking the difference between the after-tax cash flows expected upon termination of the new and old assets.

All capital budgeting decisions can be viewed as replacement decisions because expansion decisions are merely replacement decisions in which all cash flows from the old assets are zero.

Sunk Costs and Opportunity Costs

Sunk costs are cash outlays that have already been made (past outlays) and therefore have no effect on the cash flows relevant to the current decision. As a result, sunk costs should not be included in a project's incremental cash flows. If the money would not be returned if you did not go ahead with the project, then it is a sunk cost. Opportunity costs are cash flows that could be realized from the best alternative use of an owned asset. They, therefore, represent cash flows that will not be realized as a result of employing that asset in the proposed project. Because of this, any opportunity costs should be included as cash outflows when one is determining a project's incremental cash flows.

Finding the Initial Investment, Operating Cash Inflows, and Terminal Cash Flow

Finding the Initial Investment

LG 3

The initial investment in the capital budgeting procedure is found by subtracting all cash inflows occurring at time zero from all cash outflows occurring at time zero. There are three basic variables that make up the initial investment. They are (1) the installed cost of the new asset, (2) the after-tax proceeds from the sale of the old asset, and (3) the change in net working capital.

The installed cost of the new asset is found by adding the cost of the new asset to its installation cost. Installation costs are any additional costs that are necessary to place an asset into operation.

The after-tax proceeds from the sale of an old asset decrease the firm's initial investment in the new asset. These proceeds are found by subtracting applicable taxes from the old asset's sale price and should be net of any removal or cleanup costs.

When calculating the initial investment, the change in net working capital must be included. If current assets increase more than current liabilities, an increased investment in net working capital would occur and an initial *outflow* associated with the project would result. The opposite is true for a decrease in net working capital. A decrease in net working capital will result in an initial *inflow*.

Installed cost of new asset =
  Cost of new asset
  + Installation costs
– After-tax proceeds from sale of old asset =
  Proceeds from sale of old asset
  ±Tax on sale of old asset
± Change in net working capital
Initial Cash Flow

Discuss the Tax Implications Associated with the Sale of an Old Asset

To understand the tax consequences for the sale of the old asset, the concept of book value must be understood. Book value is simply the installed cost of an asset minus its accumulated depreciation. With the concept of book value in mind, it is recognized that there are several tax situations that result from the sale of an old asset. Capital gains is the sale price less the initial price. Recaptured depreciation in the initial price less the book value. If the sale price is less than the book value, then there is a capital loss, which is the book value less the sale price. These situations are summarized in the following table.

Situation	**Consequence**
The sale of the asset is for more than its initial purchase price.	Two components: (1) capital gains, and (2) recaptured depreciation. Both are taxed as ordinary income.
The sale of the asset is for more than its book value but less than its initial purchase price.	No capital gain but firm still experiences recaptured depreciation, which is taxed as ordinary income.
The sale of the asset is for its book value.	The firm breaks even, and there are no tax consequences.
The sale of the asset is for less than its book value.	This will result in a loss, and the tax effect of the loss can be used to offset ordinary income.

Finding the Operating Cash Inflows

Operating cash inflows are the relevant cash inflows that are provided by a long-term investment. Three things need to be considered when measuring the cash flows provided by a project. The cash flows that are provided by a project are (1) *incremental,* (2) *after-tax,* and (3) *cash inflows.*

The incremental cash flows represent the additional cash flows that are expected to result from a proposed project. Incremental cash flows are important because the financial manager is only concerned with how much more or less operating cash flows a new project provides.

The cash flows provided by a proposed project must be measured on an after-tax basis to ensure fairness when evaluating competing projects.

Benefits that are provided by a proposed project must be measured on a cash flow basis and not by "accounting profits." Accounting profits do not pay the firm's obligations; only cash pays the firm's liabilities.

The operating cash flows each year can be calculated using the income statement format and adding back all noncash items such as depreciation.

Finding the Terminal Cash Flow

The terminal cash flow is the relevant cash flow that is provided by the liquidation of a long-term investment at the end of its useful life. The terminal cash flow is the after-tax cash flow, separate from the operating cash inflows that occur in the final year of a project. For replacement projects, the proceeds from the sale of both the new and old assets must be considered. For expansion, renewal, and other types of projects, the proceeds from the old asset will be zero. The components of the terminal cash flow are described in the following table.

After-tax proceeds from the sale of new asset =
Proceeds from sale of new asset
±Tax on sale of new asset
After-tax proceeds from sale of old asset =
Proceeds from sale of old asset
±Tax on sale of old asset
± Change in net working capital
Terminal Cash Flow

Recognizing Risk in Capital Budgeting Projects and Discussing Risk and Cash Inflows, Scenario Analysis, and Simulation as Behavioral Approaches for Dealing with Risk

LG 4

In the discussion of capital budgeting thus far we have assumed that all investments have the same level of risk as the firm. Unfortunately, there is a great deal of uncertainty that must be dealt with when making capital budgeting decisions. This chapter introduces a number of ways to handle this uncertainty.

Behavioral Approaches for Dealing with Risk

There are several approaches for dealing with risk in capital budgeting. Here we discuss three: break-even cash flows, scenario analyses, and simulation.

Break-even cash flows are the minimum level of cash inflows necessary for a project to be acceptable. Then we could use a statistical analysis, putting a probability on the break-even cash flow.

Scenario analysis is a behavioral approach that uses a number of possible values for a given variable to assess its impact on a firm's return. One of the most common scenario approaches in capital budgeting is to estimate a project's NPV by using a pessimistic (worst), most likely (expected), and optimistic (best) cash flow. Behavioral approaches let the analyst get a "feel" for how the changing cash flows affect the project. Advanced scenario analyses evaluate the impact on returns of simultaneous changes in multiple variables.

Simulation is the process of applying predetermined probability distributions and random numbers to estimate risky outcomes. By combining the various cash flow components together in a mathematical model and repeating the process numerous times using random-number-generated probability distributions, the financial manager can simulate an NPV for projects of similar risk.

Risk Adjustment Techniques

LG 5

Risk and cash inflows require that the analyst reflect unusually high or low risk by either adjusting the discount rate used to compute the present value or by adjusting the cash flows to a level that is sure to be received. Both of these methods require subjective estimates of either the discount rate or the cash flow. The discount rate may be adjusted to match that of a firm already doing business in a field similar to what is being analyzed. Adjusting the cash flows is most appropriate if some portion of the projected revenue stream is guaranteed, say by a binding contract. When cash flows are adjusted in this manner, they should be discounted using a risk free interest rate. In this chapter we will adjust for risk by using risk-adjusted discount rates.

A common approach to risk adjustment is to adjust the discount rate to produce a risk-adjusted discount rate (RADR). The higher the risk of a project, the higher the RADR will be, and the NPV will consequently be lowered. One popular method for firms to determine the appropriate RADR is for the firm to develop some type of market risk-return function that bases discount rates on a project's coefficient of variation, a common measure of risk. By setting RADRs at different coefficients of variation, an appropriate RADR can be found for projects of varying risk.

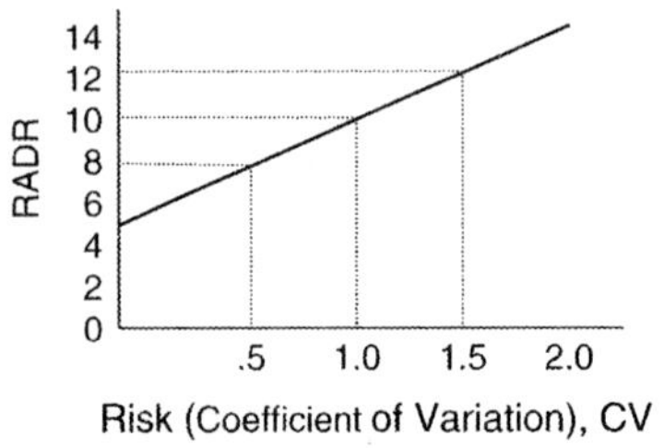

Another approach for determining RADRs would be to place projects with similar risk into risk classes and assign an appropriate RADR for each class. For example, all projects with below-average risk may be assigned an RADR of 6 percent. All projects with average risk may be assigned an RADR of 10 percent, and projects with the highest risk may be assigned an RADR of 15 percent.

Capital Budgeting Refinements—Unequal Lives, Real Options, and Capital Rationing

LG 6

Sometimes it is necessary to make adjustments in the analysis of capital budgeting projects to accommodate special circumstances. There are two areas in which special forms of analysis are used: (1) comparing mutually exclusive projects having unequal lives and (2) capital rationing by a firm.

Differing Lives

It is incorrect to compare projects with unequal lives using NPV. One popular remedy is to adjust the cash flows of each project until they are of equal length. For example, if one project was two years long and another was four years long, the short one could be doubled. This method is easy to understand but can become difficult if the projects are not multiples of each other.

A second solution to unequal lives is to convert the NPV of projects with unequal lives into equal amounts so that the projects can be compared fairly. The annualized net present value approach is a three-part process.

Step 1 Calculate the NPV of each project using the appropriate cost of capital.

Step 2 Convert the NPV into an annuity having the life of the project, n. That is, find an annuity that has the same life and the same NPV as the project.

Step 3 The project with the highest ANPV should be selected.

Example. Differing project lives

Your firm is considering the two projects detailed in the following table. Which project would you take? Note that because the projects have very different lives, either the replacement chain approach or the annualized net present value approach must be used in your analysis. Assume a 12 percent discount rate.

Year	Project 1	Project 2
Year 0	–10,000	–10,000
Year 1	3,750	2,500
Year 2	3,750	2,500
Year 3	3,750	2,500
Year 4	3,750	2,500
Year 5	3,750	2,500
Year 6		2,500
Year 7		2,500
Year 8		2,500
Year 9		2,500
Year 10		2,500

Solution

If the projects are compared ignoring the fact that they have very different lives, Project 2 is the best.

NPV_1 = \$3,518 (Calc: FV = 0, PMT = 3,570, I = 12, N = 5, CPT PV = 13,518) –10,000

NPV_2 = \$4,126 (Calc: FV = 0, PMT = 2,500, I = 12, N = 10, CPT PV = 14,126) –10,000

Because the NPV_2 is greater than NPV_1 you might be tempted to say Project 2 is the better. This conclusion ignores the idea that if Project 1 is accepted, another project could be taken at the end of five years.

You can use the equivalent annualized net present value approach. This method is shown below.

Step 1: Compute the NPV for each project—this was done above. NPV_1 = \$3,518 and NPV_2 = \$4,126.

Step 2: Compute the annualized net present value by find an annuity that has the same life and the same NPV as the project.

$ANPV_1$: CALC: PV = 3,518, N = 5, I = 12, FV = 0, CPT PMT = \$975.87

$ANPV_2$: CALC: PV = \$4,126, N = 10, I = 12, FV = 0, CPT PMT = \$730.06

Project 1 is again selected because $ANPV_1$ is greater than $ANPV_2$.

Real Options

Real options refer to opportunities that are embedded in capital projects and that enable managers to alter their cash flows and risk in a way that affects project acceptability. These are sometimes referred to as strategic options.

Some of the more common types of real options include abandonment, flexibility, growth, and time. Many of these options are embedded in capital projects and explicit recognition of their value will cause the project's strategic NPV to differ from its traditional NPV.

$$NPV_{strategic} = NPV_{traditional} + \text{Value of real options}$$

It is important to realize that the attraction of real options when determining NPV could cause an otherwise unacceptable project to become acceptable.

Capital Rationing

Firms often find that they cannot accept all of the projects that have positive net present values because of limited funds. It may be because they have self-imposed limits on borrowing or because they wish to fund their growth through internally generated funds. There are two customary techniques for dealing with limited funds: the internal rate of return (IRR) approach and the NPV approach.

The IRR approach requires that the analysts sort the available projects in order by decreasing IRR. The projects are then graphed, along with the cost of capital. Projects above the cost of capital line are accepted, and those below are rejected. The following figure demonstrates this method.

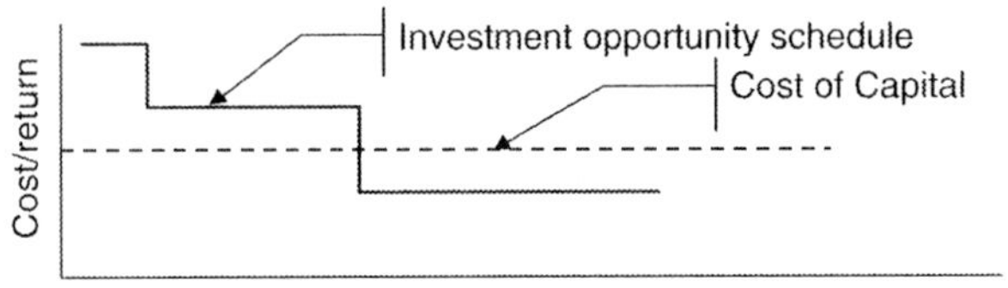

The first two projects would be accepted, and the third would be rejected. Although the IRR approach is intuitive, there is no guarantee that the projects accepted will maximize total dollar returns and, therefore, owners' wealth.

The NPV approach can be more tedious to apply but is preferred. The NPV of each project is computed. Various feasible combinations of projects are evaluated until the set of projects is identified that maximizes the total NPV of all projects.

■ Sample Problems and Solutions

Example 1.

A firm has the opportunity to replace its five-year-old popcorn machine with a brand new one costing \$10,000. The new machine will require \$5,000 in freight and installation charges. An increase in net working capital of \$2,000 is also required. The old machine was originally purchased for \$8,000 and had an expected life of 10 years. The company uses straight line depreciation. The old machine can be sold today for \$5,000. The new popcorn machine is not expected to increase revenues but will decrease operating costs from \$3,000 a year to \$1,800 per year. The new machine has a life of 10 years and will be depreciated using the straight line method. The company has a cost of capital of 10 percent, and its tax rate is 40 percent. What is the net upfront cost, and what are the annual after-tax cash flows in years 1 through 10?

Solution

Step 1: Calculate the book value of the old machine.

$$\$8{,}000/10 = \$800 \times 5 \text{ years old} = \$4{,}000$$
$$\$8{,}000 - \$4{,}000 = \$4{,}000$$

The old machine is being depreciated using straight line depreciation. To find the annual depreciation, divide the original book value by 10. Find the accumulated depreciation by multiplying by the number of years the asset has been on the books. Finally, the current book value will be the original book value minus the accumulated depreciation.

Step 2: Calculate the gain or loss on the old machine.

Sales price	\$5,000
–Book value	\$4,000
Gain on sale	\$1,000

Step 3: Calculate the tax paid on the gain, or taxes saved on the loss, from the sale of the old machine.

$$\$1{,}000 \times 0.40 = \$400$$

Because there was a gain on the sale of the old machine, taxes must be paid. The amount of the tax due is 40 percent of the gain. Had the book value been greater than the sales price, there would have been a tax savings.

Step 4: Calculate the installed cost of the new machine.

Cost	\$10,000
installation and shipping	\$ 5,000
	\$15,000

Step 5: Calculate the increase in net working capital.

\$2,000 is given in the problem.

Note: There is really no calculation needed for this problem; however, sometimes you may need to find the change in the difference between current assets and current liabilities.

Step 6: The net initial investment is the sum of Step 3, Step 4, and Step 5 minus the proceeds from the sale of the old machine. Remember that Step 3 will be negative if the old machine was sold at a loss.

+New Project Cost	–15,000
–Proceeds from old machine	+5,000
+Increase in NWC	– 2,000
(+ or –) Tax effects	– 400
Net Initial Investment	–12,400

Step 7: Calculate the net annual after-tax cash flows in years 1 through 10. Use the following format:

ΔR
$-\Delta 0$
$-\Delta D$
EBT
–tax
EAT
$+\Delta D$
CFAT

where:

$\Delta R = Rev_{new} - Rev_{old}$

$\Delta O = \text{Operating Cost}_{new} - \text{Operating Cost}_{old}$

$\Delta D = \text{Depreciation}_{new} - \text{Depreciation}_{old}$

	Years 1–5	Years 6–9	Year 10
ΔR	0	0	0
–Δ0	+1,200	+1,200	1,200
–ΔD	–700	–1,500	–1,500
EBT	+500	–300	–300
–tax	–200	+120	+120
EAT	300	–180	–180
+AD	+700	+1,500	+1500
CFAT	+1,000	+1,320	+1,320
Net working capital			+2,000
Terminal year cash flow			3,320

***Note*:** The main area of confusion on this problem is the change in the depreciation. The new depreciation is $15,000/10 = $1,500 per year. The old depreciation was $800 per year. The change in depreciation is the new minus the old = $1,500 – $800 = $700 per year. After five years, the old machine will be fully depreciated, so in years 6–10, depreciation is $1,500 per year.

Example 2. Expansion-Type Cash Flow

A firm is considering expanding its line of soft drinks as a result of market share. The new production line will cost $250,000 plus $100,000 for a new building and installation. Net working capital is expected to increase by $50,000. The new line will be depreciated over its expected seven-year life using straight line depreciation. Revenues generated by the new line are expected to be $20,000 the first year and are expected to increase at a rate of $5,000 per year thereafter. Operating costs are expected to be $10,000 in the first year and are expected to increase at a rate of $1,500 per year thereafter. After seven years, the project will have a salvage value of $10,000. The firm has a cost of capital of 12 percent and a tax rate of 40 percent. What are the net initial investment and annual cash flows associated with the project? *Hint*: The steps are very similar to Example 1 above.

Solution

Initial Investment	
Purchase price	$250,000
Installation	100,000
Initial cost	350,000
Increase in Working Capital	50,000
Net investment	$400,000

The initial cost of $350,000 is the amount that will be depreciated over the life of the project. Because the machine will have a $10,000 residual value, the annual depreciation will be ($350,000 – $10,000)/7 = **$48,571**.

	Annual Cash Flows						
	Year 1	Year 2	Year 3	Year 4	Year 5	Year 6	Year 7
Rev	$20,000	25,000	30,000	35,000	40,000	45,000	50,000
–Oper Costs	10,000	11,500	13,000	14,500	16,000	17,500	19,000
–Deprec	48,571	48,571	48,571	48,571	48,571	48,571	48,571
EBT	–38,571	–35,071	–31,571	–28,071	–24,571	–21,071	–17,571
–/+Tax	+15,429	+14,028	+12,628	+11,228	+9,828	+8,428	+7,028
EAT	–23,142	–21,043	–18,943	–16,843	–14,743	–12,643	–10,543
+Deprec	48,571	48,571	48,571	48,571	48,571	48,571	48,571
Working Capital	0	0	0	0	0	0	50,000
Salvage	0	0	0	0	0	0	10,000
Net Cash Flow	+25,429	27,528	29,628	31,728	33,828	35,928	98,028

Example 3. Initial Investment Calculation

Columbia Corporation is considering replacing its existing packaging machine, which was purchased three years ago at a cost of $275,000. The machine can be sold today for $120,000. It is being depreciated using MACRS and a five-year recovery period (the depreciation schedule is shown below). A new machine will cost $350,000 to purchase and install. The new machine will require an increase in accounts receivable of $50,000, an increase in inventory of $30,000, and an increase in accounts payable of $20,000. Assume a 40 percent tax rate. Calculate the initial investment associated with the replacement of the machine.

Recovery Year	Percentage by Recovery Year 5 Years
1	20%
2	32%
3	19%
4	12%
5	12%
6	5%
Totals	100%

Solution

The book value of the old machine is $275,000 – (0.2 + 0.32 + 0.19) × $275,000 = $79,750.

The gain/loss of the sale of the old machine today: $120,000 – $79,750 = $40,250.

Taxes on the sale: $40,250 × 0.4 = $16,100.

After-tax proceeds from the sale: $120,000 – $16,100 = $103,900.

Change in net working capital: Increase in A/R + Increase in Inventory – Increase in A/P = $50,000 + $30,000 – $20,000 = $60,000.

Initial investment: cost of the new machine – after-tax proceeds from the sale of the old machine – Change in NWC = \$350,000 – \$103,900 – \$60,000 = \$186,100.

Example 4. Breakeven Analysis

Batmaker Inc. is considering purchasing a new bat-making machine. The cost of the machine is \$30,000, and the company expects the new machine will produce a steady income for its five-year life. If Batmaker requires a 13 percent return on its investment, what minimum yearly inflow will be necessary for the company to go forward with this project?

Solution

Solve for the breakeven cash flow: N = 5, I = 13, PV = 30000, FV = 0 and CPT PMT = 8529.44. So the company must have a yearly income of at least \$8,529.44 to go forward with the project.

Example 5. Risk Adjusted Discount Rates

Hockey Unlimited is considering investing in one of two mutually exclusive projects, A and B. The firm's cost of capital, r, is 14 percent, and the risk-free rate, R_F, is 3 percent. The firm has gathered the basic cash flow and risk index data for each project, as show in the following table:

	Projects (j)	
	A	B
Initial investment (CF_0)	\$18,000	\$20,000
Year (t)	Cash Flows (CF_t)	
1	\$7,500	\$5,000
2	7,500	7,000
3	7,500	9,000
4	7,500	13,000
Risk index (RI)	1.2	0.6

a. Find the net present value of each project using the firm's cost of capital. Which project is preferred?

b. The firm uses the following equation to determine the risk-adjusted discount rate, $RADR_j$, for each project, j:

$$RADR_j = R_F + [RI_j \times (r - R_F)]$$

where

R_F = risk-free rate of return
RI_j = risk index for project j
r = cost of capital

Use the RADR for each project to determine its risk-adjusted NPV. Which project is preferable in this situation?

c. Compare and discuss your findings in part a and b. Which project do you recommend that the firm accept?

Solution

a. Project A

$N = 4, I = 14, PMT = 7{,}500, FV = 0, CPT\ PV = 21{,}852.84$

$NPV = 21{,}852.84 - 18{,}000 = 3{,}852.84$

Project B

$CF_0 = -20{,}000; CF_1 = 5{,}000; CF_2 = 7{,}000; CF_3 = 9{,}000; CF_4 = 13{,}000$

Set I = 14, Solve for NPV = 3,544.02

So project A is preferred because it has the greater NPV.

b. $RADR_A = 3\% + 1.2 \times (14\% - 3\%) = 17.2\ \%$

$RADR_B = 3\% + 0.6 \times (14\% - 3\%) = 9.6\%$

Project A

$N = 4, I = 17.2, PMT = 7{,}500, FV = 0, CPT\ PV = 20493.46$

$NPV = 20{,}493.46 - 18{,}000 = 2{,}493.46$

Project B

$CF_0 = -20{,}000; CF_1 = 5{,}000; CF_2 = 7{,}000; CF_3 = 9{,}000; CF_4 = 13{,}000$

Set I = 9.6, Solve for NPV = 6,235.12

So project B is preferred because it has the greater NPV.

c. After adjusting the discount rate, even though all projects are still acceptable, the ranking changes. Project B has the highest RADR NPV and should be chosen.

Example 6. Unequal Lives—ANPV Approach

Coin Machines Inc. is considering two alternative coin counting machines. Machine A has an expected life of four years, will cost $11,500, and will produce net cash flows of $5,000 per year. Machine B has a life of eight years, will cost $12,000, and will produce net cash flow of $3,500 per year. Coin plans to operate the machines for eight years. Inflation is expected to be zero, and the company's cost of capital is expected to be 13 percent. Which new machine should it use?

Solution

Machine A

$N = 4, I = 13, PMT = 5{,}000, FV = 0, CPT\ PV = 14{,}872.36$

$NPV = 14{,}872.36 - 11{,}500 = 3{,}372.36$

Machine B

$N = 8, I = 13, PMT = 3{,}500, FV = 0, CPT\ PV = 16{,}795.70$

$NPV = 16{,}795.70 - 12{,}000 = 4{,}795.70$

Based on NPV, Machine B should be chosen. However, the machines have unequal lives, so the ANPV should be found.

Machine A

PV = 3,372.36, N = 4, I = 13, FV = 0, CPT (ANPV) PMT = 1,133.76

Machine B

PV = 4,795.70, N = 8, I = 13, FV = 0, CPT (ANPV) PMT = 999.36

So based on ANPV, Machine A should be chosen. This is the method that should be used with mutually exclusive projects with unequal lives.

Example 7. Capital Rationing—IRR and NPV Approaches

Ilex Corporation is attempting to select the best of a group of independent projects competing for the firm's fixed capital budget of $850,000. The cost of capital is 15 percent. The firm has summarized the key data to be used in selecting the best group of projects:

Project	Initial investment	IRR	NPV @ 15 %
A	$200,000	18%	$95,000
B	400,000	20%	100,000
C	350,000	19%	90,000
D	250,000	17%	75,000
E	150,000	16%	20,000
F	200,000	15%	0

Which projects should the firm implement and why? (Show both the IRR method and the NPV method)

Solution

Ranking by IRR:

Project	Initial investment	IRR	Total investment
B	$400,000	20%	$400,000
C	350,000	19%	750,000
A	200,000	18%	950,000

Only projects B and C could be invested in because adding project A would go over the budget, and there are no other projects for $100,000 or less. The total net present value from projects B and C is $190,000.

Ranking by NPV:

The best option would be to invest in projects A, B, and D, which uses the whole $850,000 budget and generates a total net present value of $270,000, which is $80,000 greater than the IRR method.

Example 8. Risk-Adjusted Project

Suppose that you are evaluating a project with a cost of $15,000 and five annual cash inflows in $6,500 each. The return on the market is 12 percent, the risk-free rate of interest is 6 percent, and the firm's beta is 1.2. The project you are evaluating is very different from those usually undertaken by the firm. Another firm, which is exclusively in the business of doing projects like the one under consideration, has a beta of 2.0. Should you recommend that the firm accept the project? Assume the firm is all equity financed.

Solution

$r_i = 6\% + 2 \times (12\% - 6\%) = 18\%$.

Yes because it has an *NPV* of $5,327 [PMT = 6,500, N = 5, I = 18, FV = 0, PMT PV = 20,327; 20,327 – 15,000 = 5,327 = NPV).

Example 9. Real Options

Erickson Corporation is considering abandoning a product line with no debt. The line could be sold for $300,000 after taxes, or it could be kept, and it will produce after-tax cash flows of $130,000 for each of three years. In addition, the possibility of modernizing the line with after-tax cash consequences solely for the modernization are as follows:

–$60,000	$30,000	$40,000	$20,000
0	1	2	3

Should Erickson keep, abandon, or modernize if the discount rate is 11 percent?

Solution

NPV for keeping: PMT = 130,000, N = 3, FV = 0, I = 11, PMT PV = 317,683; NPV = 317,683 – 300,000 = 17,683

NPV for modernizing: CFO = –360,000, C01 = 160,000, F01 = 1, C02 = 170,000, F02 = 1, C03 = 150,000, F03 = 1, I = 11, NPV = $31,799

Because the highest NPV is for modernizing, Erickson should modernize.

■ Quick Drill Problems

These problems are quick and easy practice to see how well you have understood the concepts in this chapter. The answers follow.

1. Your company has decided to purchase a new computer system for $150,000. Installation and training will cost an additional $40,000 and $25,000, respectively. What is the net purchase price?

2. From Question 1, what amount would you use as the basis for computing depreciation on the new computer system?

3. The computer system your company is currently using was purchased three years ago for $140,000 and has a MACRS depreciable life of five years. What is the book value of the old computer system?

4. Your company is selling the old computer system for $25,000. Using the information from Question 3, what is the taxable gain (loss) on the sale of the old computer system? Assuming a 40 percent tax rate, what is the tax on the sale? Is this amount added to or subtracted from the net purchase price when computing the initial investment?

5. Using the information from Questions 1 through 4, what is the initial cash flow, assuming the computer system will reduce net working capital by $32,000.

6. Using the straight-line depreciation method, what is the annual depreciation of commercial property that cost $2.5 million? Assume a 39-year depreciable life.

7. Suppose you owned a sporting goods store and decided to expand your operations by opening a new store. You project that revenue will reach $250,000 for the new store in the first year and increase by 20 percent over the next three years. Expenses are estimated at 75 percent of sales, the initial cost of the expansion was $400,000 (depreciable over 39 years using the straight-line method), and your tax rate is 40 percent. What are the annual cash flows from 2001 to 2004?

8. As the manager of a printing company, you were impressed at a local trade show with a next-generation printing machine. It occupies less space, is more efficient, and provides a better quality product. This machine costs $260,000 with annual operating costs of $15,000. However, two years ago you purchased a printing machine for $190,000 with annual operating costs of $29,000. You are depreciating the old printing machine over three years using MACRS, as you would for the new printing machine. Revenues will increase from $760,000 to $790,000 with the replacement of the old printing machine. Note that the change in operating cost is $15,000 – $29,000. Assume a 40 percent tax rate and compute the annual cash flows for the next four years.

9. Using the information from Question 8, you plan to sell the new printing machine after four years for $60,000. What is the terminal cash flow?

10. The initial cash flow for a project is –$250 million. Cash flows are projected to be –$150 million in 2001, $300 million in 2002, $450 million in 2003, and $500 million in 2004. If the cost of capital is 12 percent and additional costs in 2004 are –$250 million, what is the NPV?

11. Suppose the initial cash flow for a project is –$500 million. Cash flows are projected to be –$300 million and 2001, –$200 million in 2002, $250 million in 2003, $500 million in 2004, $600 million in 2005, and $850 million in 2006. If the cost of capital is 14 percent and additional costs for terminating the project in 2006 are –$175 million, what is the NPV?

12. Suppose the cash inflow in the last year of a project is $22 million. The new asset is to be sold for $5 million, resulting in a tax of $450,000. If net working capital was initially decreased by $400,000, what is the terminal cash flow?

Answers

1. –$215,000
2. –$215,000
3. $40,600
4. –$15,600 (loss); –$6,240; added
5. –$151,760
6. $64,103
7. $41,603; $49,103; $58,103; $68,903
8. $49,320, $67,880, $42,000, $33,680
9. $69,680
10. $334,410,223
11. –$166,873,735
12. $26,150,000

Long Sample Problem

The calculation of cash flows for capital budgeting is one of the more complex activities you will face in this course. These problems are often very long and complex. The following is a long problem, and its cash flow solution as solved using Excel.

■ Cash Flow Analysis

Prepare your answer using an Excel spreadsheet.

You have decided that your town needs a new Mexican restaurant. It will feature live music, appetizers, and a well-stocked bar. You are trying to decide whether you should invest in this venture. Your first step is to prepare a complete capital budgeting analysis for the five years the restaurant will operate before you sell it.

After talking with local vendors, restaurant owners, bankers, and builders you collect the following data.

You will need to make tenant improvements to the property. These will include cooking equipment, a stage, seating, and interior décor. The construction is estimated to cost $1,500,000. An additional $275,000 will be spent on chairs, tables, bar equipment, and decorations. Depreciation will be over seven years using MACRS. You determine that you will require an average cash balance of $15,000 and inventory of $20,000. Accounts payable should average $10,000. You learn that you can borrow the money to pay for these expenses at a 15 percent interest rate.

You feel that your chances of success will increase if you have a market study performed by a group of ECU faculty. The charge for this report will be $100,000.

The site you have selected for the lounge is in a building currently owned by your family. Your parents have said that you can use the retail space in any way you wish for free. After checking on local lease rates, you determine this space would lease for $75,000 per year. Your family also owns another restaurant downtown. You predict that your new restaurant will decrease the old restaurant's revenues by $15,000 per year. Your parents tell you that this sum will be taken from your annual family stipend.

Revenues are estimated to be $500,000 the first year. Revenues are expected to increase by 20 percent the second year, 15 percent the third year, and to continue increasing at 10 percent thereafter. Fixed annual operating costs are estimated to be as follows: employee salaries = $110,000; heat, electric, water, janitorial = $75,000. The food and liquor bill is expected to be 15 percent of revenues. Total taxes are estimated to be 40 percent of net revenues.

Your plan is to run the bar for five years, then to sell it to an investor for $2,000,000.

Prepare a cash flow analysis that includes the initial investment, the annual cash flows, and the terminal cash flow.

Solution

Mexican Restaurant

Initial cash flow	
Purchase price	
Construction	(1,500,000)
Appointments	(275,000)
	(1,775,000)
NWC	(25,000)
Net Cost	(1,800,000)

		Growth rate of revenue	20%	15%	10%	10%
		Annual Cash Flows				
		Year 1	**Year 2**	**Year 3**	**Year 4**	**Year 5**
Revenues		500,000	600,000	690,000	759,000	834,900
Expenses						
Salaries		110,000	110,000	110,000	110,000	110,000
Utilities		75,000	75,000	75,000	75,000	75,000
Food and Liquor		75,000	90,000	103,500	113,850	125,235
Rent (opportunity cost)		75,000	75,000	75,000	75,000	5,000
Indirect expense		15,000	15,000	15,000	15,000	15,000
Total Operating Expenses		350,000	365,000	378,500	388,850	400,235
Depreciation (7 year)		(248,500)	(443,750)	(301,750)	(230,750)	(159,750)
Total depreciation		(248,500)	(443,750)	(301,750)	(230,750)	(159,750)
Earnings before tax		(98,500)	(208,750)	9,750	139,400	274,915
Tax (40%)		39,400	83,500	(3,900)	(55,760)	(109,966)
Earnings after-tax		(59,100)	(125,250)	5,850	83,640	164,949
Depreciation		248,500	443,750	301,750	230,750	159,750
Sale of lounge						2,000,000
Tax on sale						(643,800)
NWC	Initial Invest					25,000
Net Cash Flow	(1,800,000)	189,400	318,500	307,600	314,390	1,705,899

Tax Calculation

The calculation of the tax on the sale of the assets in the terminal year can be confusing.

Accumulated Depreciation = Sum of depreciation	(1,384,500)
Book Value = Initial cost – Accumulated depreciation	390,500
Taxable gain = Sales price – Book value	1,609,500
Tax on gain = 0.4	$1,609,500

■ Study Tips

1. Building cash flow estimates is a job ideally suited to the spreadsheet. Try to use Excel or Lotus to work out your homework assignments. The copy command and summation commands make the process go quickly. You will find it much easier to correct your mistakes as well.

2. Review, if necessary, finding book value. You will need this ability when working any replacement-type cash flow problem. Determine the accumulated depreciation, then subtract this figure from the original book value.

3. An easy check you can perform is to make sure that the earnings before tax (EBT) and the tax figure always have different signs. For example, if earnings before tax is –$17,571, the tax will be a *positive* amount. Likewise, if the EBT is positive, the tax will be negative.

4. Some students wonder why they need to subtract out depreciation when computing EBT, since depreciation is added back in at the bottom. The reason is that by subtracting out depreciation, an accurate EBT can be found for estimating taxes.

5. Keep in mind that financial managers focus on actual cash flows. Any operating cash flows that arise from depreciation write-offs that would have supposedly occurred if an asset had not been sold is not included directly in the analysis. The taxable portion of the gain arising from an earlier asset sale is lower because the it has a larger non-depreciated basis.

6. One of the most significant points made by this chapter is that you must adjust for risk when appropriate. Pay particular attention to when risk adjustment may be required.

7. The reason for discussing breakeven analysis in this chapter is that it highlights how increasing fixed costs increases risk by increasing breakeven sales.

8. Remember that the adjustment of the discount rate is preferred to any method of subjectively adjusting cash flow for risk.

9. In the last chapter we noted that IRR leads to different rankings than does NPV. In this chapter IRR is shown as a ranking tool. Keep in mind that when IRR is used to rank, the results should be confirmed by using NPV.

■ Student Notes

■ Sample Exam—Chapter 11

True/False

T F 1. An outlay for advertising and management consulting is considered a current expenditure.

T F 2. A firm's fixed asset investments provide the basis for the firm's earning power and value.

T F 3. If a firm has unlimited funds to invest, then all the mutually exclusive projects that meet its minimum investment criteria will be implemented.

T F 4. The purchase of additional machinery or a new factory is an example of the renewal type of capital expenditure.

T F 5. The book value of an asset is equal to the asset's after-tax proceeds that are provided after the asset has been sold.

T F 6. Projects should be evaluated on the basis of accounting profits because these profits actually cover the company's obligations.

T F 7. The relevant cash flows for a proposed project are the incremental after-tax cash outflows and the resulting cash inflows.

T F 8. All projects will have an initial investment, cash inflows, and a terminal cash flow.

T F 9. When considering international projects, it is important to consider exchange rate risk and political risk.

T F 10. Net working capital is the amount by which a firm's current assets exceed its current liabilities.

T F 11. When the sale of an asset is equal to its book value, a firm will have to pay taxes on recaptured depreciation.

T F 12. The salvage value of an asset should be net of any removal or cleanup costs associated with the termination of a project.

T F 13. A capital expenditure is a short-term outlay of funds that is expected to produce cash flows usually in less than one year.

T F 14. Sunk costs are cash flows that could be realized from the best alternative use of an owned asset.

T F 15. Opportunity costs are cash outlays that have already been made and therefore have no effect on the cash flows relevant to the current decision.

T F 16. It is appropriate to use the firm's cost of capital when evaluating any capital budgeting project.

T F 17. If a project is more risky than those the firm usually accepts, it will have to have a greater NPV to be acceptable.

T F 18. Behavior approaches to evaluating capital projects allow the analyst to get a feel for the level of project risk.

T F 19. Scenario analysis is an approach that evaluates the impact of simultaneous changes in a number of variables.

T F 20. A net present value profile is a graphical presentation of the NPV at various discount rates.

T F 21. In reference to capital budgeting, risk is the chance that a project has a high degree of variability in the initial investment.

T F 22. The break-even cash inflow is the minimum level of cash inflow necessary for a project to be acceptable.

T F 23. A higher risk adjusted discount rate (RADR) will result in a higher NPV.

T F 24. When dealing with mutually exclusive, unequal-lived projects, the use of NPV to select the better project may result in an incorrect decision.

T F 25. Most firms practice capital rationing, which means that projects will be competing for limited dollars.

T F 26. Simulation is a statistics-based behavioral approach that applies predetermined probability distributions and random numbers to estimate risky outcomes.

T F 27. Real options are opportunities that are embedded in capital projects that enable managers to alter their cash flows and risk in a way that affects project acceptability.

Multiple Choice

1. The most common motive for adding fixed assets to a firm is
   a. replacement.
   b. expansion.
   c. renewal.
   d. government regulation.

2. A series of equal annual cash flows is termed
   a. a mixed stream.
   b. nonconventional.
   c. conventional.
   d. an annuity.

3. The relevant cash flows of a project are best described as
   a. accounting cash flows.
   b. incremental cash flows.
   c. incidental cash flows.
   d. accounting profits.

4. An important cash inflow associated with a replacement project is
   a. the cost of the new asset.
   b. installation costs.
   c. the sale of the old asset.
   d. the tax savings.

5. The change in net working capital must be considered a part of the _________ when evaluating a capital budgeting project.
   a. initial cash outflow
   b. initial investment
   c. operating cash inflow
   d. terminal cash flow

6. The sale of an ordinary asset for its book value results in
   a. no tax benefit.
   b. recaptured depreciation.
   c. an ordinary tax benefit.
   d. a capital gain.

7. One of the basic techniques used to evaluate after-tax operating cash flows is to
   a. add cash expenses to net income.
   b. subtract depreciation from net income.
   c. add noncash charges to net income.
   d. subtract retained earnings from net income.

8. A corporation has decided to replace an existing machine with a newer model. The old machine had an initial purchase price of $30,000 with $5,000 installation costs and has $20,000 in accumulated depreciation. If the 40 percent tax rate applies to the corporation and the old asset can be sold for $10,000, what will be the tax effect of the replacement?
   a. loss of $2,000
   b. no effect
   c. benefit of $2,000
   d. loss of $4,000

9. The cash flows in an expansion project are different from those in a replacement project in that the cash flows from the old asset in an expansion project
   a. will be evaluated on an after-tax basis.
   b. will always be zero.
   c. will always be negative.
   d. will always equal the terminal cash flow.

10. The book value of an asset is equal to the
   a. purchase price minus depreciation expense.
   c. purchase price minus accumulated depreciation.
   b. purchase price minus recaptured depreciation.
   d. fair market value.

11. Using the following data, what is the change in net working capital for Upgrade Corporation?

Current Account	Change in Balance
Cash	+$12,000
Accounts Payable	+$21,000
Accruals	+$ 6,000
Depreciation	+$10,000
Inventories	+$24,000

a. +$9,000
b. –$1,000
c. –$4,000
d. +$19,000

12. The __________ evaluates capital expenditure projects to determine whether they meet a firm's minimum acceptance criterion.

a. ranking approach
b. book value
c. P/E multiple approach
d. accept-reject approach

13. The initial cash flows and operating cash inflows are referred to as the

a. necessary cash flows.
b. ordinary cash flows.
c. relevant cash flows.
d. consistent cash flows.

14. Giancarlo has received an inheritance from his rich uncle and is contemplating the purchase of a Suzuki XL7. In an attempt to make a rational decision, Giancarlo has identified the following cash flow estimates:

Negotiated price of new Suzuki XL7:	$24,675
Taxes and fees on new car purchase:	$ 1,732
Proceeds from the trade-in of old car:	$ 9,285
Estimated value of Suzuki XL7 in five years:	$ 7,285
Estimated value of old car in five years:	$ 3,572
Estimated annual repair cost on Suzuki XL7:	$ 350
Estimated annual repair cost on old car:	$ 925

What would be Giancarlo's initial investment in the Suzuki XL7?

a. $13,370
b. $15,390
c. $17,122
d. $26,406

15. What would be Gioncarlo's operating cash flow in Year 5?
    a. ($575)
    b. $350
    c. $375
    d. $4,127

16. What would be Gioncarlo's terminal cash flow?
    a. $3,177
    b. $3,572
    c. $3,752
    d. $7,324

17. A firm acquired a machine two years ago for $50,000. Its current book value is $35,000, and it can be sold today for $55,000. The firm's capital gain is ______, and its recaptured depreciation is _____.
    a. $20,000; $35,000
    b. $5,000; $20,000
    c. $5,000; $15,000
    d. $15,000; $5000

18. In a replacement problem, the old machine was two years old at the time of the decision and was being depreciated over five years at $15,000, and the new machine was being depreciated over five years for $20,000 per year. What is the change in depreciation in year 3 and in year 4?
    a. $5,000 and $20,000, respectively
    b. $5,000 and $5,000, respectively
    c. $5,000 and $0, respectively
    d. $5,000and $15,000, respectively

19. A method for evaluating a project that uses a number of possible values for a given variable, such as cash inflows, to assess its impact on the firm's return is
    a. sensitivity analysis.
    b. scenario analysis.
    c. certainty equivalent.
    d. risk-adjusted cash flow.

20. A statistically based behavioral approach to project analysis that applies predetermined probability distributions is the
    a. simulation method.
    b. scenario approach.
    c. sensitivity approach.
    d. certainty equivalent.

21. Which of the following is not a normal risk consideration unique to multinational firms?
    a. exchange rate risk
    b. political risk
    c. tax issues related to transfer pricing
    d. market risk

22. A firm that has limited dollars available for capital expenditures often practices
    a. capital dependency.
    b. net working capital constraints.
    c. capital rationing.
    d. bond financing.

23. A firm has unlimited funds in which to evaluate the following five projects.

Project	**Status**	**Return(%)**
1	Independent	13
2	Independent	12
3	Independent	15
4	Mutually Exclusive	14
5	Mutually Exclusive	16

    What will the ranking of the projects be on the basis of their returns, from the best to the worst?
    a. 5, 3, 1, 4, and 2
    b. 3, 2, 1, 5, and 4
    c. 3, 2, 1, 4, and 5
    d. 5, 3, 4, 1, and 2

24. The rate of return that must be earned on a given project to compensate owners for projects with unusual risk levels is the
    a. WACCAT.
    b. risk-adjusted discount rate.
    c. IRR.
    d. risky interest rate.

25. From a practical standpoint, it makes more sense to adjust for risk in capital budgeting using
    a. portfolio interest rates.
    b. certainty equivalents.
    c. WACCAT.
    d. risk-adjusted discount rates.

26. After adjusting the following cash flow for unequal lives using the replacement chain approach, the NPV of Project 1 is _________. and the NPV of Project 2 is _________, where the required return is 12 percent.

	Project 1	**Project 2**
Year 0	–10,000	–10,000
Year 1	3,750	2,500
Year 2	3,750	2,500
Year 3	3,750	2,500
Year 4	3,750	2,500
Year 5	3,750	2,500
Year 6		2,500
Year 7		2,500
Year 8		2,500
Year 9		2,500
Year 10		2,500

a. 3,518; 4,125
b. 4,125; 5,514
c. 5,514; 4,125
d. none of the above

27. Unlike the IRR criteria, the NPV approach assumes an interest rate equal to the___________ when not adjusting for risk.

a. market interest rate
b. project's internal rate of return
c. risk-free rate of return
d. firm's cost of capital

28. The objective of ________ is to select the group of projects that provides the highest overall net present value and does not require more dollars than are budgeted.

a. scenario analysis
c. capital rationing
b. simulation
d. sensitivity analysis

29. The following two projects are mutually exclusive. Both have a cost of capital of 10 percent.

	Project A	**Project B**
Length of cash inflows	5	7
NPV	$12,000	$14,000

What is the annualized net present value of Project A and Project B?

a. $3,166 and $2,876
b. $2,378 and $1,850
c. $2,986 and $4,197
d. $4,174 and $3,915

30. A project that has a coefficient of variation of zero is considered
    a. slightly risky.
    b. a bad investment.
    c. very risky.
    d. risk free.

31. An increase in the risk adjusted discount rate will result in
    a. no change to the NPV.
    b. a decrease in the NPV.
    c. an increase in the NPV.
    d. an increase in the IRR.

32. The amount by which the required discount rate exceeds the risk-free rate is called the
    a. risk equivalent.
    b. risk premium.
    c. excess risk.
    d. market risk function.

33. Vang Vu has carefully investigated investment in Mung Technologies. Purchasing 100 shares would cost $5,000. However, Mung Technologies is expected to offer an annual dividend of $60 and be sold at a price of $6,500 in three years. Although Vang normally requires a return of 10 percent, this company has a low beta, resulting in a required return of only 8.5 percent. Should Vang invest in Mung Technologies?
    a. No, because the net present value is $242.
    b. Yes, because the required rate of return is less than that normally required.
    c. Yes, because the net present value is $89.
    d. Yes, because the net present value is $242.

34. Risk classes and risk-adjusted discount rates are based upon
    a. the CAPM, but not risk-return behavior of past projects.
    b. government-dictated classifications and yields.
    c. risk-return behavior of past projects but not the CAPM.
    d. risk-return behavior of past projects and the CAPM.

35. Tim Demmer, a rodeo clown, is evaluating the cost of a recreation vehicle (RV), which costs $98,000. Tim plans on keeping the RV for five years, at which time he believes he will be able to sell it for $55,000. Using the RV will cut down on hotel and meal costs by about $16,000 per year, on an after-tax basis. Assuming a discount rate of 7 percent for projects of this relative low risk, what is the RV's ANPV?
    a. $1,037
    b. $1,119
    c. $1,663
    d. $6,817

36. Sarah Alain has to make a choice between two investments. Investment A has a net present value of $55,000, lasts six years, and requires a discount rate of 9 percent. Investment B has a net present value of $45,000, lasts five years, and has a discount rate of 13 percent. Project selection should be based upon
    a. annualized net present value.
    b. discount rates.
    c. initial costs.
    d. net present value.

37. In the previous problem, which investment should Sarah Alain select?
    a. Investment A, which has an annualized net present value of $9,167.
    b. Investment A, which has an annualized net present value of $12,261.
    c. Investment B, which has an annualized net present value of $9,000
    d. Investment B, which has an annualized net present value of $12,794.

38. West Flooring evaluated a proposed capital expenditure for new floor sanding equipment and found the NPV to be –3,000. West decided to assess the possibility of real options embedded in the firm's cash flows. It was determined the project could be abandoned at the end of two years, resulting in and additional NPV of $3,500. What is the strategic NPV?
    a. $6,500
    b. –$500
    c. $500
    d. –$3,000

39. Referring to problem 38, if instead of abandoning the project, West estimated an NPV from merging with East Flooring in two years to be $5,500 and an NPV of sharing the equipment with North during the second year to be $1,500. Management felt there would be a 60 percent chance of the merging and a 25 percent chance of the equipment sharing. What is the value of the real options, and what is the strategic NPV for the proposed capital expenditure project?
    a. $675 and $3,675, respectively
    b. $3,675 and $675, respectively
    c. $6,675 and $3,000, respectively
    d. $3,000 and –$675, respectively

Essay

A firm is contemplating replacing its old copier with a new one that is faster and easier to operate. The old machine was being depreciated over five years using straight line depreciation. Its original installed cost was $12,000. The old machine has been in use four years. It can be traded in for $2,500. The new machine will cost $21,000. It will also be depreciated over five years using the straight line method. It is not expected to have a residual value. Net working capital will decrease because supply levels can be reduced by $1,000.

Although the new machine is not expected to affect revenues, it will result in labor savings of $5,000 per year due to its greater speed. Reduced training expenses are expected to save an additional $1,500 per year. The firm is in the 40 percent tax bracket.

1. What is the initial cash flow for the copier?

2. What is the operating cash flow?

3. What is the terminal cash flow?

4. As a top manager, you must decide which of the proposed projects should be accepted for the upcoming year because only $6 million is available for the next year's capital budget. What is the total NPV of the projects that should be accepted?

Project	Cost (millions)	NPV (millions)	PI
A	3.25	0.80	1.25
B	1.75	0.52	1.30
C	4.50	0.69	1.15
D	3.75	0.95	1.25
E	1.25	0.25	1.20
F	0.50	0.25	1.50

5. Your company must decide between two positive NPV projects with unequal lives. The three-year project has an initial cash flow of –$2 million, with cash inflows of $4 million per year. The six-year project has an initial cash flow of –$6 million, with cash inflows of $3 million per year. Assume a cost of capital of 12 percent and that neither project has additional end of project costs. What is the NPV for each project using the equal annual annuity method? Which project should your company accept?

6. Discuss the differences between the IRR and NPV method to capital rationing. Which is better and why?

■ Chapter 11 Answer Sheet

True/False

1. F
2. T
3. F
4. F
5. F
6. F
7. T
8. F
9. T
10. T
11. F
12. T
13. F
14. F
15. F
16. F
17. F
18. T
19. T
20. T
21. F
22. F
23. F
24. T
25. T
26. T
27. T

Multiple Choice

1. B
2. D
3. B
4. C
5. B
6. A
7. C
8. C $30,000 – 5,000 = 25,000
   $25,000 – 20,000 = 5,000
   $10,000 – 5,000 = 5,000
   $5,000 × 0.40 = $2,000
9. B
10. C
11. A ($12,000 + 24,000) – ($21,000 + 6,000) = $9,000
12. D
13. C
14. C
15. A
16. C
    Cap gain = 55,000 – 50,000,
    Recap. Dep = 50,000 – 35,000
17. D
18. A
    Year 3:20,000 – 15,000 = 5,000;
    Year 4: 20,000 – 0 = 20,000
19. B
20. A
21. D
22. C
23. C
24. B
25. D
26. C
    NPV_I = (Calc: PMT = 3750, N = 10, I = 12, FV = 0, CPT PV = 21188) – (Calc: PMT = 0, N = 5, FV = 10,000, I = 12, CPT PV = 5,674) – $10,000
    NPV = $21,188.25 – $5,674 – $10,000
    NPV = $5,514
27. D
28. C
29. A ($ANPV_A$: Calc: PV = 12000, N = 5, I = 10, FV = 0,CPT PMT = 3166), ($ANPV_B$: Calc: PV = 14,000, N = 7, I = 10, FV = 0, CPT PMT = 2,876)
30. D
31. B
32. B
33. D
34. D
35. C
36. A
37. D
38. B (–3,000 + 3,500 = 500)
39. C [(0.6 × 5,500) + (0.25×1,500) = 3,675;
    –3,000 + 3,675 = 675]

Essay

1.

Initial Cash Flow	
Purchase price	+21,000
Sale of old copier	–2,500
Tax on sale of old machine	+40
Decrease in net working capital	–1,000
Total	17,540

2. & 3.

	Operating Cash Flows				
	Year 1	**Year 2**	**Year 3**	**Year 4**	**Year 5**
Rev	0	0	0	0	0
Oper	6,500	6,500	6,500	6,500	6,500
Depr	1,800	4,200	4,200	4,200	4,200
EBT	4,700	2,300	2,300	2,300	2,300
Tax	1,880	920	920	920	920
EAT	2,820	1,380	1,380	1,380	1,380
Depr	1,800	4,200	4,200	4,200	4,200
Working Cap	0	0	0	0	(1,000)
Net Cash Flow	4,620	5,580	5,580	5,580	4,580

4. F, B & D; total cost = $6 million; total NPV = $1.72 million

5. Three year = $26,394,184; six year = $12,838,714; accept three year

6. The IRR approach and the NPV approach to capital rationing both involve ranking projects on the basis of IRRs. Using the IRR approach, a cut-off rate and a budget constraint are imposed. The NPV first ranks projects by IRR and then takes into account the present value of the benefits from each project in order to determine the combination with the highest overall net present value within the budget constraint. The benefit of the NPV approach is that it guarantees a maximum dollar return to the firm, whereas the IRR approach does not; therefore, the NPV approach is better.

Chapter 12
Leverage and Capital Structure

■ Chapter Summary

The chapter discusses how a firm manager may decide between debt or equity financing. The term "capital structure" refers to how much long-term financing comes from debt and how much from equity. Increasing debt increases fixed-cost financing or financial leverage. Increasing the amount of long-term assets or fixed costs increases operating leverage. No definitive methods are presented that will lead the manager to choose a particular capital structure because finance has simply not found any model that gives us a reliable answer. This chapter does discuss the issues that must be considered, however. It also provides some tools to use when deciding how to finance a particular project. A key to understanding this chapter is that fixed costs are stated in terms of dollars per period, while variable costs are in dollars-per-unit terms.

LG 1 *Discuss leverage, capital structure, breakeven analysis, the operating breakeven point, and the effect of changing cost on the breakeven point.* Leverage results from the use of fixed costs to magnify returns to a firm's owners. Capital structure, the firm's mix of long-term debt and equity, affects leverage and therefore the firm's value. Breakeven analysis is useful for determining both the level of sales necessary to cover all operating costs and the profitability at different sales levels. Breakeven analysis can also be used to look at the impact on profits from varying the different cost components.

LG 2 *Understand operating, financial, and total leverage and the relationships among them.* Operating leverage refers to leverage due to the use of fixed operating costs by the firm to magnify the effects of changes in sales on EBIT. Financial leverage is the use of fixed-cost financing such as debt and preferred stock to magnify the effects of changes in EBIT on EPS. Total leverage relates the level of earnings to the use of fixed operating costs and fixed financing costs. Total leverage is a combination of operating leverage and financial leverage. Financial leverage is multiplied by operating leverage to compute total leverage.

LG 3 *Describe the types of capital, external assessment of capital structure, the capital structure of non-U.S. firms, and capital structure theory.* Capital structure refers to how the firm is financed through the combination of debt and equity. Ratios such as the debt ratio and the times-interest-earned ratio are used to assess the ability of the firm to manage fixed cost financing. Early capital structure theories focused on the tax deductibility of debt to conclude that firms should be 100 percent debt financed. More recent theories note bankruptcy costs and agency costs mitigate the value of the tax deductibility of interest expense.

LG 4 *Explain the optimal capital structure using a graphical view of the firm's cost-of-capital functions and a zero-growth valuation model.* The optimal capital structure is the combination of debt and equity that maximizes the value of the firm. This will occur at the point where the tax benefit of the debt just equals the cost of potential bankruptcy and agency cost. The zero-growth valuation model defines the firm's value as its net operating profits after taxes (NOPAT), or after-tax EBIT, divided by its weighted average cost-of-capital. Assuming NOPAT is constant, the value of the firm is maximized by minimizing its weighted-average cost-of-capital (WACC).

LG 5 *Discuss the EBIT-EPS approach to capital structure.* The EBIT-EPS approach evaluates capital structures in light of the returns they provide the firm's owners and their degree of financial risk. Under the EBIT-EPS approach, the preferred capital structure is the one that is expected to provide maximum EPS over the firm's expected range of EBIT. Graphically, this approach reflects risk in terms of the financial breakeven point and the slope of the capital structure line. The major shortcoming of EBIT-EPS analysis is it concentrates on maximizing earnings (returns) rather than owners' wealth, which considers risk as well as return.

LG 6 *Review the return and risk of alternative capital structures, their linkage to market value, and other important considerations related to capital structure.* Keep focused on the idea that the optimal capital structure is one that optimizes firm value. The optimal capital structure will balance return and risk in a way that maximizes market value.

■ Chapter Notes

Leverage

Leverage results from the use of fixed operating costs and/or fixed financing cost to magnify returns to the firm's stockholders. By fixed costs we mean costs that do not rise or fall with changes in a firm's sales. With an increase in leverage, the risk and return to the firm usually increases. There are three basic types of leverage:

Operating leverage—the relationship between the firm's sales revenue and its earnings before interest and taxes (EBIT).

Financial leverage—the relationship between the firm's EBIT and its common stock earnings per share (EPS).

Total leverage—the relationship between the firm's sales revenues and its EPS.

Breakeven Analysis

LG 1 Before examining the different types of leverage, it is important to understand the effects of fixed costs on the firm's operations. Breakeven analysis is used to determine the level of operations necessary to cover all operating costs and to evaluate the profitability associated with various levels of sales. To determine the operating breakeven point (the level of sales necessary to cover all operating costs), the EBIT must be set equal to zero. In other words, what level of sales must be achieved to just cover all costs and therefore have an EBIT of zero?

The following formula is used to calculate EBIT:

$$\text{EBIT} = (P \times Q) - FC - (VC \times Q)$$

where:

P = sales price per unit

Q = sales quantity in units

FC = fixed operating costs per period

VC = variable operating costs per period

By setting the EBIT equal to zero, the breakeven point in units can be determined. By rearranging and solving for Q, the following breakeven equation is derived:

$$Q = \frac{FC}{P - VC}$$

The breakeven point can be represented graphically as follows:

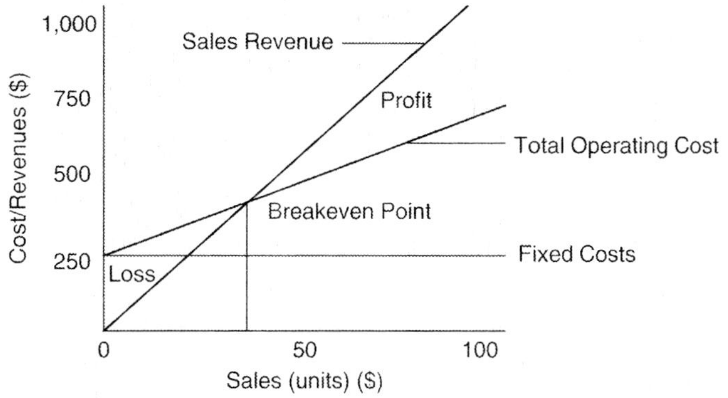

The point where total operating cost equals sales revenue is where EBIT equals zero. This point is the firm's operating breakeven point. If fixed costs increased, total operating costs would also increase, and therefore, the level of sales necessary to break even would increase.

Operating Leverage

LG 2

Operating leverage reflects the use of a firm's fixed operating costs to generate changes in the firm's EBIT through increased sales. Because fixed costs by their very nature do not fluctuate with sales, when sales increase, total operating costs would increase but only by the increase in variable costs. Therefore, firms that choose to use more fixed costs relative to variable costs can magnify their earnings. To measure the degree of operating leverage (DOL), the following equation should be used:

$$\text{DOL} = \frac{\%\text{ change in EBIT}}{\%\text{ change in sales}}$$

Financial Leverage

Financial leverage reflects the use of fixed financial costs to magnify the effects of changes in EBIT on the firm's EPS. Fixed financial costs are those associated with the use of debt and preferred stock because interest and preferred dividends are fixed and must be paid. The degree of financial leverage (DFL) can be measured by using the following formula:

$$\text{DFL} = \frac{\%\text{ change in EPS}}{\%\text{ change in EBIT}}$$

Total Leverage

Total leverage reflects the use of a firm's fixed costs, both operating and financial, to magnify the effect of changes in sales on the firm's EPS. The degree of total leverage (DTL) can be measured by using the following formula:

$$\text{DTL} = \frac{\%\text{ change in EPS}}{\%\text{ change in sales}}$$

or

$$\text{DTL} = \text{DOL} \times \text{DFL}$$

High operating leverage and high financial leverage will cause total leverage to be high. The opposite is also true, but the relationship between operating leverage and financial leverage is multiplicative instead of additive.

Example

A firm has a DOL of 1.4 and a DFL of 4.5. What is the firm's DTL?

$$\text{DTL} = \text{DOL} \times \text{DFL}$$

$$\text{DTL} = 1.4 \times 4.5 = 6.3$$

The Firm's Capital Structure

LG 3

The term "capital" refers to the long-term funds of the firm. There are two types of capital: (1) debt, and (2) equity.

Debt capital includes all the long-term borrowing of the firm. Debt is cheaper than the other forms of financing due to three risk-reduction factors:

1. A higher priority claim against earnings or assets available for payment
2. Stronger legal pressure against the company to make payment over preferred or common stockholders
3. The significantly lower debt costs to the firm due to the tax deductibility of interest payments.

Equity capital includes all long-term funds provided by the firm's owners, the stockholders. There are two forms of equity capital: (1) preferred stock, and (2) common stock equity, which includes common stock and retained earnings. The order of cost for equity, from most expensive to least expensive, is common stock, retained earnings, and preferred stock.

Often by examining a selection of financial ratios, a firm's financial leverage can be assessed. The debt ratio measures the degree of indebtedness of the firm. The higher this ratio, the higher the firm's financial leverage. The times-interest-earned-ratio measures the firm's ability to meet interest payments as they come due. In general, a poor times-interest-earned-ratio is associated with a high degree of financial leverage.

Capital structure decisions are very important to the firm. Poor capital structure decisions can result in a high cost of capital, resulting in less acceptable NPV projects; whereas effective capital structure decisions can lower the cost of capital, resulting in more acceptable NPV projects, thereby increasing the value of the firm.

Capital Structure Theory

LG 4

At this point in time there is not a specific methodology that can be used to determine a firm's optimal capital structure. However, there is theoretical and empirical research that suggests, by assuming perfect markets for securities (equity), that there is a theoretical optimal capital structure based on balancing the benefits and costs of debt financing. The cost of debt financing is based on the following factors:

Tax benefits—The interest paid on debt is tax deductible, thus reducing the firm's cost of debt.

Increased probability of bankruptcy—The probability a firm will go bankrupt is based on both business risk and financial risk. Business risk varies from firm to firm and is not affected by capital structure decisions. The firm's capital structure directly affects its financial risk. The more fixed-cost financing—debt and preferred stock—a firm has in its capital structure, the greater its financial leverage and risk, and the greater the chance of bankruptcy.

Agency cost—This is the cost of the lender's monitoring and controlling the firm's actions that are centered on the debt capital.

Asymmetric information—This situation arises when managers of a firm have more information about operations and future prospects than do investors.

The optimal capital structure is the point at which the weighted average cost of capital is minimized, thereby maximizing the firm's value. This can be represented by the following equation in which EBIT is held constant and the WACC, r_a, is minimized to provide the highest possible firm value, V.

$$V = \frac{\text{EBIT} \times (1 - T)}{r_a}$$

The EBIT-EPS Approach to Capital Structure

LG 5

The EBIT-EPS approach to capital structure involves selecting the capital structure that maximizes earnings per share over the expected range of earnings before interest and taxes (EBIT). The EBIT-EPS approach can best be presented graphically and involves plotting the EPS for various expected EBITs of different capital structures.

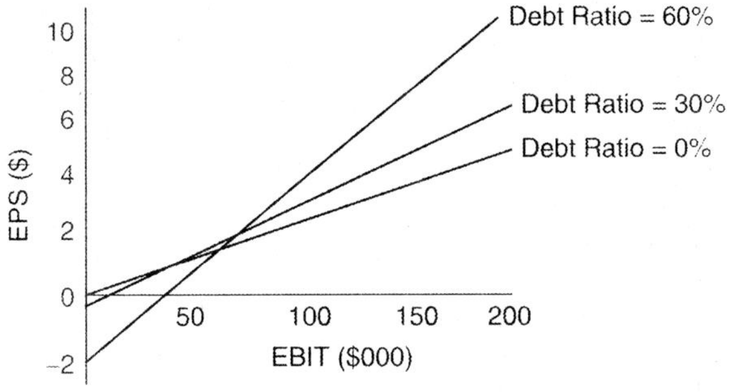

The graphical representation of the EBIT-EPS approach reveals that over certain ranges of expected EBIT, an optimal capital structure exists that maximizes EPS. The graph also reveals that the steeper the slope of the capital structure line, the greater the financial risk of that capital structure. The increased risk is caused by increased leverage.

The basic shortcoming of the EBIT-EPS approach to capital structure is that it concentrates on earnings rather than owners' wealth maximization. The EBIT-EPS approach generally ignores risk, which plays a major role in determining owners' wealth.

Choosing the Optimal Capital Structure

LG 6

An optimal capital structure is one that balances risk and return in a way that maximizes owners' wealth. To link these two factors, the risk associated with each capital structure should be included in the required rate of return for that structure. This can be done by estimating the beta for each alternative capital structure and using the CAPM to calculate the required return, r_s.

Another approach would involve linking the financial risk associated with each capital structure directly to the required return. An example of this would involve estimating the required return for different levels of risk as measured by the coefficient of variation of EPS. Once a capital structure has been set and an EPS has been determined, the per-share value of the firm can be found by using the zero-growth valuation model. We can use this model with the simplifying assumption that all earnings are paid out as dividends.

$$P_o = \frac{EPS}{r_s}$$

There is one important point to remember when evaluating various capital structures. The maximization of profit does not necessarily mean owners' wealth maximization. The financial manager should only implement capital structures that increase the wealth of the firm's shareholders.

Sample Problems and Solutions

Example 1. Degree of Operating Leverage

Given the data below, compute the degree of operating leverage.

	2014	2013
Sales (dollars)	$5,000	$10,000
Less variable operating cost	2,500	5,000
Less fixed operating cost	2,500	2,500
EBIT	$ 0	$ 2,500

Solution

The formula for the degree of operating leverage is:

$$DOL = \frac{\%\text{ change in EBIT}}{\%\text{ change in sales}}$$

so

$$\%\Delta EBIT = \frac{0 - 2,500}{2,500} = -100\%$$

$$\%\Delta SALES = \frac{5,000 - 10,000}{10,000} = -50\%$$

so

$$DOL = \frac{-100\%}{-50\%} = 2$$

Example 2. Degree of Financial Leverage

Using the data in the following table, compute the degree of financial leverage.

	2014	2013
EBIT	$6,000	$10,000
Less interest	2,000	2,000
Earnings before tax	4,000	8,000
Less tax	1,600	3,200
Earnings after tax	2,400	4,800
Less preferred dividend	2,400	2,400
Earnings for shareholders	0	2,400
Earnings per share (EPS)	$ 0	$ 2.40

Solution

The degree of financial leverage is found using the following equation:

$$DFL = \frac{\%\text{ change in EPS}}{\%\text{ change in EBIT}}$$

so

$$\%\Delta EPS = \frac{0 - 2.4}{2.4} = -100\%$$

$$\%\Delta EBIT = \frac{6,000 - 10,000}{10,000} = -40\%$$

and

$$DFL = \frac{-100\%}{-40\%} = 2.5$$

Example 3. Degree of Total Leverage

Use the data presented in Examples 1 and 2 to compute the degree of total leverage (DTL).

Solution

The degree of total leverage is computed as the product of DOL and DFL.

$$\text{DTL} = \text{DOL} \times \text{DFL}$$

Using the values for DOL and DFL already computed,

$$\text{DTL} = 2 \times 2.5$$
$$\text{DTL} = 5$$

Example 4. Breakeven Analysis

Debbie's Deli Delight is a New York-style delicatessen that specializes in massive roast beef sandwiches and spicy hot dogs. Her average sale is $5.00. She has fixed costs of $2,500 per month, which includes her rent, utilities, and payments on her equipment. Her variable costs are $1.75 per sandwich for the food and labor. She is open 6 days per week. Compute her breakeven sales per month.

Solution

The formula for computing the breakeven is:

$$Q = \frac{FC}{P - VC}$$

where Q is the quantity of goods needed to break even, FC is the fixed costs, VC is the variable cost/unit, and P is the average selling price/unit.

$$Q = \frac{\$2{,}500}{\$5 - \$1.75}$$
$$Q = \frac{\$2{,}500}{\$3.25} = 769.23 \text{ sandwiches}$$

The breakeven point in sales is found by multiplying the number of sandwiches that must be sold times the price per sandwich.

$$\text{BE sales} = 769.23 \times \$5 = \$3{,}846.15$$

Example 5. Breakeven Analysis

Mary's Meatballs by Mail Inc. distributes authentic Italian meatballs by mail order. Her only fixed costs are a telephone line and the printing and mailing of a catalog. These costs total $300 per month. Her average order is $5. The variable cost per sale is $2.50, which includes postage. What are Mary's breakeven sales per month?

Solution

Using the same method as above, the quantity breakeven point is:

$$Q = \$300/(\$5 - \$2.50) = 120 \text{ meatball orders.}$$

The breakeven point in dollars is $\$5 \times 120 = \$600$.

Note that Mary's Meatballs has much lower fixed costs than Debbie's Deli; Mary's also enjoys lower risk because of a much lower breakeven point. On the other hand, if sales were to skyrocket, Debbie's stands to earn more because her variable costs are lower.

■ Study Tips

1. Recognize that there is no one solution to the capital structure question. Because different firms have different bankruptcy and agency costs, different levels of debt will serve to maximize firm value in different industries.

2. One more time, the goal of optimizing capital structure is to maximize the value of the firm.

3. There are three ways to measure the leverage of your firm. The DOL measures your use of fixed-cost assets. Breakeven analysis demonstrates how risk increases as the use of fixed assets increase. DFL increases as the use of debt increases. EBIT-EPS analysis can help choose between financing options that have different levels of debt. Finally, DTL is the product of DOL and DFL. Firms are believed to want to try to keep DTL from getting too high. So if one type of leverage is high, the firm will keep the other low. For example, if a firm must have many fixed assets to produce its product, it will try to keep its debt level low.

■ Student Notes

■ Sample Exam—Chapter 12

True/False

T F 1. The firm's operating breakeven point is the level of sales necessary to cover all of the firm's fixed costs.

T F 2. The amount of leverage in the firm's capital structure can significantly affect its value by affecting return and risk.

T F 3. An increase in the sales price per unit will generally reduce the firm's breakeven point.

T F 4. Leverage results from the use of fixed-cost assets or funds to magnify returns to the firm's owners.

T F 5. Operating leverage is concerned with the relationship between the firm's sales revenue and its operating expenses.

T F 6. Financial leverage is concerned with the relationship between the firm's earnings after interest and taxes and its common stock earnings per share.

T F 7. In theory, there is an optimal capital structure.

T F 8. Total leverage is determined by adding the firm's operating leverage to its financial leverage.

T F 9. Whenever the percentage change in EBIT resulting from a given percentage change in sales is greater than the percentage change in sales, financial leverage exists.

T F 10. A shift toward less fixed cost decreases business risk, which in turn causes EBIT to increase by more for a given increase in sales.

T F 11. The probability that a firm will go bankrupt is largely dependent on its business risk and financial risk.

T F 12. Assuming that EBIT is held constant, the value of the firm is maximized when the WACC is maximized.

T F 13. The EBIT-EPS approach tends to concentrate on maximization of owners' wealth rather than earnings maximization.

T F 14. The steeper the slope of the line relating EBIT to EPS, the greater the financial risk.

T F 15. If expected EBIT is less than the EBIT-EPS indifferent point computed for two financing plans, the plan with greater debt should be chosen.

T F 16. Capital structure is the mix of long-term debt and equity maintained by the firm.

T F 17. The theoretical optimal capital structure is based on balancing the benefits and costs of interest payments to be deducted in calculating taxable income.

T F 18. The costs of debt financing results from (1) increase probability of bankruptcy, (2) agency costs, and (3) asymmetric information.

Multiple Choice

1. _________ analysis is a technique used to assess the returns associated with various cost structures and levels of sales.
   a. Risk
   b. Breakeven
   c. Ratio
   d. Return

2. Fixed costs are a function of _________, not sales, and are typically contractual.
   a. EBIT
   b. variable cost
   c. time
   d. marginal cost

3. At the operating breakeven point, _________ equals zero.
   a. fixed cost
   b. sales revenue
   c. variable cost
   d. EBIT

4. If a firm's fixed operating costs increase, the firm's operating breakeven point will
   a. increase.
   b. not be affected.
   c. decrease.
   d. remain unchanged.

5. A firm has fixed operating costs of $4,500, the sale price per unit of its product is $12, and its variable cost per unit is $7. The firm's breakeven point in units is
   a. 375.
   b. 6,000.
   c. 643.
   d. 900.

6. _________ results from the use of fixed-cost assets or funds to magnify returns to the firm's owners.
   a. Debt
   b. Equity
   c. Leverage
   d. Operating margin

7. Through the effects of financial leverage, when EBIT decreases, earnings per share will
   a. decrease.
   b. increase.
   c. remain unchanged.
   d. not be affected.

8. A decrease in fixed operating costs will result in _________ in the degree of operating leverage.
    a. a decrease
    b. no change
    c. an increase
    d. an undetermined change

9. The basic sources of capital for a firm include all of the following EXCEPT
    a. preferred stock.
    b. dividends.
    c. long-term debt.
    d. retained earnings.

10. _________ risk is the risk of being unable to cover operating costs.
    a. Total
    b. Leverage
    c. Financial
    d. Business

11. _________ leverage is concerned with the relationship between sales revenues and earnings before interest and taxes.
    a. Financial
    b. Total
    c. Operating
    d. Fixed

12. The inexpensive nature of long-term debt in a firm's capital structure is due to the fact that
    a. interest payments are tax deductible.
    b. it is riskier than other forms of capital.
    c. debt is based on the risk-free rate.
    d. equity capital has a fixed return.

13. The probability of bankruptcy is determined by
    a. total risk.
    b. market risk.
    c. financial risk.
    d. operating risk.

14. Fixed financing costs include
    a. interest expense.
    b. interest and depreciation expense.
    c. interest expense and preferred stock dividends.
    d. interest expense, preferred dividends, and common stock dividends.

15. The key differences between debt and equity capital include all of the following EXCEPT
    a. maturity.
    b. tax treatment.
    c. voice in management.
    d. effect on operating leverage.

16. _________ leverage is concerned with the relationship between earnings before interest and taxes and earnings per share.
    a. Total
    b. Financial
    c. Operating
    d. Fixed

17. A firm has fixed operating costs of $200,000, a sales price per unit of $75, and a variable cost per unit of $50. The firm's operating breakeven point in dollars is
    a. $600,000.
    b. $750,000.
    c. $8,000.
    d. $45,000.

18. The firm's degree of total leverage is found by _________ the firm's degree of financial leverage _________ its operating leverage.
    a. adding; to
    b. subtracting; from
    c. dividing; by
    d. multiplying; by

19. The basic shortcoming of the EBIT-EPS approach to capital structure is
    a. that it is difficult to compute.
    b. its disregard for the firm's dividend payout ratio.
    c. that it concentrates on EPS maximization instead of owner's wealth maximization.
    d. its disregard for the firm's retained earnings.

20. In theory, the firm should maintain financial leverage consistent with a capital structure that
    a. maximizes EPS.
    b. maximizes owner's wealth.
    c. minimizes EPS.
    d. minimizes taxes.

21. Continuing with the example in Chapter 12 where Rick Polo analyzed the benefits of a fuel savings device, let's assume that the demand for the fuel savings devices push their installment cost up to $300 and monthly fees to $17 per month. Given that the benefit of having these devices stays at $28, what is the breakeven point (in months)?
    a. 10.7 months
    b. 17.6 months
    c. 27.3 months
    d. This is a trick question; Rick would never break even.

22. Les Peila currently owns a vehicle used to transport his son's Little League team to games around the state. He currently has disposable income of $3,000, of which $500 goes to the bank in payment of the loan obtained to purchase the vehicle. When working overtime in the winter, to build up a savings account for payment of gasoline and food expenses when the team plays away from home, Les can increase his disposable income to $3,750. What is Les's degree of financial leverage?
    a. 1.20
    b. 1.25
    c. 1.30
    d. 6.00

23. The Rossins are applying for a mortgage loan. The family's monthly gross income is $5,750 and they currently have monthly installment loan obligations of $160. The $250,000 mortgage loan they are applying for will require monthly payments of $1,900. The lender requires (1) the monthly mortgage payment to be less than 27 percent of monthly gross income and (2) total monthly installment payments (including the mortgage payment) to be less than 36 percent of monthly gross income. Will the Rossins get the home loan?
    a. No, they fail to qualify based on both ratios.
    b. No, because their monthly mortgage will be 33.0 percent of gross income.
    c. No, because their monthly installment loan obligation will be 35.8 percent of gross income.
    d. Yes, the Rossins qualify on the basis of both ratios.

24. Assume Quick Music Download Inc. has a sales price of $1.25 per unit, variable cost of $0.50 per unit and fixed operating costs of $4000. Their current sales are 10,000 units. What is their degree of operating leverage?
    a. 2.00
    b. 2.14
    c. 1.75
    d. 0.9

25. Refer to problem 24. If Quick Music Download has no preferred stock but has interest payments of $2000. What is their degree of total leverage?
    a. 2.00
    b. 2.14
    c. 1.36
    d. 5.00

Essay

1. What is the general relationship between the operating leverage, financial leverage, and total leverage of the firm?

2. What are some of the factors that influence the use of debt in the capital structure?

Chapter 12 Answer Sheet

True/False

1. F
2. T
3. T
4. T
5. F
6. F
7. T
8. F
9. F
10. F
11. T
12. F
13. F
14. T
15. F
16. T
17. T
18. T

Multiple Choice

1. B
2. C
3. D
4. A — $Q = \dfrac{\$4,500}{12-7} = 900 \text{ units}$
5. D
6. C
7. A
8. A
9. B
10. D
11. C
12. A
13. A
14. C
15. D
16. B — $Q_U = \dfrac{\$200,000}{75-50} = 8,000 \text{ units}$; $Q_P = 8,000 \times \$75 = \$600,000$
17. A
18. D
19. C
20. B
21. C
22. A
23. B
24. B
25. C

DOL= [Q × (P – VC)]/[Q × (P – VC) – FC]
= [10,000 × (1.25 – 0.5)] / [10,000 × (1.25 –0.5) – 4000]
= 2.14

DTL= [Q × (P – VC)]/[Q × (P × VC) × I – FC]
= [10,000 × (1.25 – 0.5)] / [10,000 × (1.25 – 0.5) – 2000 – 4000]
= 5.0

Essay

1. Operating leverage deals with the use of fixed assets. Financial leverage deals with the use of fixed-cost funds. Total leverage is the product of the two. Another way of looking at DOL is that it represents the leverage from the top of the income statement to EBIT. DFL is the leverage from EBIT to the bottom of the income statement. DTL is the leverage from the top to the bottom of the income statement.

2. Using debt as a source of some portion of the firm's long-term funds is influenced by the tax deductibility of the interest expense. But the use of debt also increases the risk that the firm will be unable to pay its fixed financial costs. Also, lenders impose restrictions on the firm, and the costs of monitoring these constraints are called agency costs. Asymmetric information may also influence the use of debt, according to research.

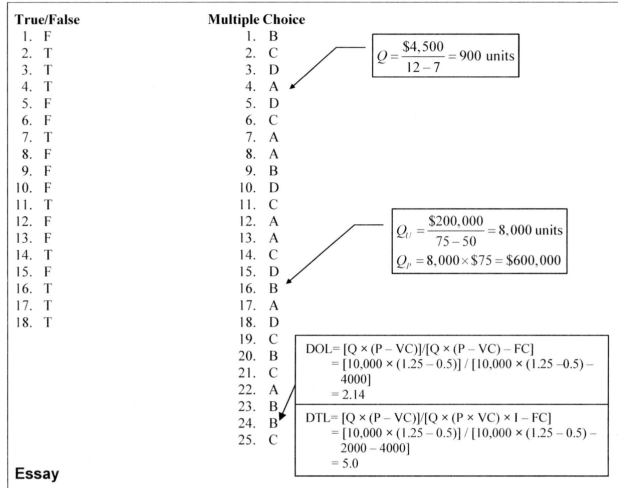

Chapter 13
Payout Policy

■ Chapter Summary

This chapter describes payouts and payout policies. Cash payouts or dividends are basically a source of cash for shareholders and provide them with information about the firm's current and future expectations. This chapter is mostly descriptive in nature. Carefully review the types of dividends and the related policies.

LG 1 *Understand cash payout procedures, their tax treatment, and the role of dividend reinvestment plans*. Cash payouts decisions are made normally each quarter by the board of directors, along with the establishment of record and payment dates. Dividends are usually paid either in cash or by offering additional shares of stock. For most individuals, the dividend tax rate is the same as on capital gains. Some firms offer dividend reinvestment plans that allow stockholders to acquire shares in lieu of cash dividends.

LG 2 *Describe the residual theory of dividends and the key arguments with regard to dividend irrelevance and relevance*. Residual theory states dividends are the residual earnings left after undertaking all acceptable investment opportunities. There is substantial debate about whether dividend policies really affect the firm's value.

LG 3 *Discuss the key factors involved in establishing a dividend policy*. Legal, contractual, and internal constraints affect a firm's payout policy by placing restrictions on the amount of cash that is available to be paid. Growth prospects, owner considerations, and market considerations also can affect dividend policy.

LG 4 *Review and evaluate the three basic types of dividend policies*. A constant-payout ratio requires that the firm pay out a fixed percentage of its earnings as dividends every year. This will result in fluctuating income to stockholders. A regular dividend policy fixes the amount of the dividend at a constant amount that is not changed from year to year unless earnings warrant a change. This provides a level payment to stockholders but may put a burden on the firm in lean years. The low-regular-and-extra dividend policy is similar to the regular dividend policy, except that its pays an extra dividend when the firm's earnings are higher than normal.

LG 5 *Evaluate stock dividends from accounting, shareholder, and company points of view*. The use of stock dividends merely shifts funds between equity accounts. The shareholders really receive nothing of value because the percentage ownership of the firm does not change. In reality, stock dividends are more costly for the firm, but they satisfy shareholders without spending cash.

LG 6 *Explain stock splits and the firm's motivation for undertaking them*. Stock dividends do not cost the firm any cash, yet tend to be popular with shareholders. Stock splits keep the price of the stock where management wants it. Stock repurchases may be the best investment available for a firm if its stock is underpriced. Also, reducing the number of shares outstanding increases earnings per share and the market price of the shares. Stock repurchases also defer the tax payments of stockholders.

Chapter Notes

Payout Policy Basics

LG 1

The term payout policy refers to the decisions firms make about whether to distribute cash to shareholders, how much cash to distribute, and by what means the cash should be distributed. The two primary ways in which a firm can distribute cash to shareholders is through a *dividend payment* and *share repurchases*. Cash payout decisions normally are made each quarter by the board of directors, along with the establishment of record and payment dates. Dividends are usually paid either in cash or by offering additional shares of stock. Since the jobs and Growth Tax Relief Reconciliation Act of 2003, for most individuals, the dividend tax rate is the same as that on capital gains. Because retained earnings—earnings that are not distributed as a form of dividends—are a form of internal financing, the payout decision can affect the firm's external financing requirements. Rapidly growing firms generally need internal financing to grow and therefore often do not pay dividends.

Dividends are divided by earnings in computation of the *dividend payout ratio*. Because dividends are more stable than earnings, the dividend payout ratio increases in a recession. During economic expansions earnings growth will exceed dividend growth, and the dividend payout ratio will drop.

Every person whose name is recorded as a stockholder on the *date of record* will at a specified future time receive a declared dividend. Because of the time needed to make bookkeeping entries, the stock will begin selling without the right to receive the current dividend four days prior to the date of record. During this time, the stock is said to be selling *ex dividend.*

The *payment date* is the actual date on which the firm will mail the dividend payment to the holder of record. The payment date is generally set a few weeks after the record date. For example:

Declaration date	Ex-date	Record date	Payment date
June 10	June 25	July 1	August 1

If the board of directors declares a dividend on June 10 with a record date of July 1 and a payment date of August 1, the ex-date will be June 25. This is four days plus a two-day weekend before the record date. Purchasers of the stock on June 24 will receive the dividend. Investors who purchase the stock on June 25 or later will not receive the dividend.

Dividend reinvestment plans (DRIPs) enable stockholders to use dividends received on the firm's stock to acquire additional shares, even fractional shares, at little or no transaction cost.

Relevance of Dividend Policy

LG 2

Residual theory of dividends: This theory states that the dividend paid by a firm should be the amount left over after all acceptable investment opportunities have been undertaken. The residual theory implies retained earnings should be used to finance all acceptable investments. If the amount of financing needed is in excess of retained earnings, new forms of financing should be used. If retained earnings are in excess of the firm's financing needs, the residual funds should be distributed to the owners of the firm in the form of dividends. Because investment opportunities would tend to vary year to year, this approach would not lead to a stable dividend. This view of dividends tends to suggest the required return of investors, r_s, is not influenced by the firm's dividend policy, a premise that suggests dividend policy is irrelevant and does not affect firm value.

Dividend irrelevance theory: This theory, put forth by Miller and Modigliani, states that in a perfect world the value of a firm is unaffected by the distribution of dividends and is determined solely by the earning power and risk of its assets. Miller and Modigliani argue that there can be a clientele effect in which different types of investors are attracted to firms with different payout policies due to tax effects but still do not change the value of the firm.

Dividend relevance theory: Gordon and Lintner theorize that stockholders prefer current dividends and there is a direct relationship between the dividend policy of the firm and its market value. Fundamental to this proposition is the bird-in-the-hand argument. They argue that current dividend payments reduce uncertainty, causing investors to require a lower return, resulting in a higher stock value.

Studies on dividends have lead to the *information content* argument that dividends act as a signal to investors about the future of a firm. An increase in dividends would be viewed as a positive signal that would cause investors to bid up the share price.

The *agency cost theory* says a firm that commits to paying dividends is reassuring shareholders that managers will not waste their money. Given this reassurance, investors will pay higher prices for firms that promise regular dividend payments.

Factors Affecting Dividend Policy

LG 3

Legal constraints: Most states prohibit corporations from paying out as cash dividends any portion of the firm's legal capital. Certain states measure legal capital by the par value of outstanding common stock, while others include any paid-in capital in excess of par.

Contractual constraints: Often a firm's ability to pay cash dividends is constrained by certain restrictive provisions in a loan agreement.

Internal constraints: A firm's ability to pay dividends is generally constrained by the amount of excess cash available to pay the dividends.

Growth prospects: Firms that are in a growth stage generally need all the funds that they can get to finance capital expenditures and usually pay little or no dividends.

Owner considerations: Firms must establish a policy that has a favorable effect on the wealth of the majority of the owners.

Market considerations: Firms must be aware of the market's expected response to types of dividend policies.

Types of Dividend Policies

LG 4

A firm's dividend policy must be formulated with two objectives in mind: maximizing the wealth of the firm's owners and providing for sufficient financing. Three of the more commonly used dividend policies are the constant-payout ratio, regular, and low-regular-and-extra dividend policy.

- *Constant-payout ratio*: A firm's dividend payout ratio is calculated by dividing the firm's cash dividend per share by its earnings per share, which indicates the percentage of each dollar earned that is distributed to the owners in the form of cash. With a constant-payout ratio, the firm pays this percentage out of earnings to the owners in the form of a dividend each period.
- *Regular dividend policy*: Firms using a regular dividend policy pay out a fixed-dollar dividend in each period. The regular dividend does not fluctuate with earnings, and dividends rarely decrease. A regular dividend policy removes owner uncertainty and reduces the risk of the stock.

- *Low-regular-and-extra dividend policy*: Some firms pay a low regular dividend and supplement this low dividend when earnings warrant it. The dividend that is paid in addition to the regular dividend is called an extra dividend. The use of an extra dividend is especially common in firms that experience cyclical shifts in earnings.

While the constant-payout ratio policy results in dividend variability and owner uncertainty, the regular dividend policy and low-regular-and-extra dividend policy reduce owner uncertainty by fulfilling their dividend expectation each period. Both the regular dividend and the low-regular-and-extra dividend policies provide good signals to investors.

Other Forms of Dividends

LG 5

Stock dividends: A stock dividend is the payment to existing owners of a dividend in the form of stock instead of cash. Therefore, the payment of a stock dividend is a shifting of funds between capital accounts rather than a use of funds. The shareholder receiving a stock dividend receives nothing of value. After the dividend is paid, the per-share value of the shareholder's stock will decrease in proportion to the dividend in such a way that the market value of their holdings in the firm will remain unchanged.

Stock splits: Stock splits are not a form of dividends, but they have the same effect on a firm's share price as that of a stock dividend. A stock split is used to lower the market price of a firm's stock by increasing the number of shares belonging to each shareholder. Most splits are in the form of a 2-for-1 split, in which two new shares are exchanged for each old share, or a 3-for-2 split, in which three new shares are exchanged for two old shares. A stock split has no effect on the firm's capital structure, but simply lowers the stock's market price to enhance trading activity. Theoretically, a stock split has no impact on investor wealth and is typically nontaxable.

LG 6

Stock repurchases: Motives for stock repurchases include obtaining shares to be used in acquisitions, having shares available for employee stock option plans, or retiring shares. Recently, stock repurchases have been used to enhance shareholder value. The enhancement of shareholder value through a stock repurchase can be achieved by (1) reducing the number of shares outstanding and thereby raising earnings per share (EPS), (2) sending a positive signal to investors in the marketplace that management believes that the stock is undervalued, and (3) providing a temporary floor for the stock price, which may have been declining.

■ Sample Problems and Solutions

Example 1. Dividend Policy

Starjump's Coffee has had the following earnings per share over the 2005–2013 period:

Year	Earnings per Share
2005	1.50
2006	1.00
2007	0.80
2008	1.20
2009	1.30
2010	1.60
2011	1.10
2012	0.90
2013	1.60

a. If the firm's dividend policy is based on a constant-payout ratio of 50 percent for all years with positive earnings, determine the annual dividend paid in each year.

b. If the firm has a regular dividend policy of paying $0.60 per share per year, regardless of the per-share earnings, until the per-share earnings remain above $1.20 for 2 periods, at which time the dividend would be increased to $0.80 per share, determine the annual dividend paid in each year.

c. If the firm's policy is to pay out $0.20 per share each period, except in those periods when earnings are above $1.00, when they pay out an extra dividend equal to 30 percent of the earnings above $1.00, determine the amount of regular and extra dividends paid each year.

Solution

a. If the firm paid out 50 percent of each year's earnings in years with positive earnings, the dividends paid per share (DPS) would be as follows:

Year	2005	2006	2007	2008	2009	2010	2011	2013	2014
DPS	$0.75	$0.50	$0.40	$0.60	$0.65	$0.80	$0.55	$0.45	$0.80

b. If the firm paid out a fixed $0.60 per share until earnings were above $1.20 for 2 years, at which time they raised the dividend to $0.80 per share, the dividend paid would be as follows:

Year	2005	2006	2007	2008	2009	2010	2011	2013	2014
DPS	$0.60	$0.60	$0.60	$0.60	$0.60	$0.80	$0.80	$0.80	$0.80

c. If the firm pays a regular dividend plus an extra dividend the payout would be as follows:

Year	2005	2006	2007	2008	2009	2010	2011	2013	2014
RDPS	$0.20	$0.20	$0.20	$0.20	$0.20	$0.20	$0.20	$0.20	$0.20
EDPS	$0.15	$0.00	$0.00	$0.06	$0.09	$0.18	$0.03	$0.00	$0.18
TDPS	$0.35	$0.20	$0.20	$0.26	$0.29	$0.38	$0.23	$0.20	$0.38

Note: RDPS = regular dividend per share, EDPS = extra DPS, and TDPS = total DPS.

Study Tips

1. This is another low-math, high-memorization chapter. On separate paper, see if you can list the different types of dividend policies, the dividend theories, and the factors that affect dividend policy.

2. Study the advantages and disadvantages of the different dividend policies carefully. This material makes good essay questions on exams.

Student Notes

■ Sample Exam—Chapter 13

True/False

T F 1. Dividend reinvestment plans (DRPs) are used by firms to allow existing shareholders to purchase additional shares of stock with their dividends.

T F 2. A stock split commonly increases the number of shares outstanding and the stock's per-share par value.

T F 3. The date of record for dividends is the actual date on which the company will mail the dividend payment to the holders of record.

T F 4. The ex dividend date period begins four business days prior to the record date.

T F 5. The dividend decision can significantly affect the firm's share price and external financing requirements.

T F 6. The residual theory of dividends is based on the argument that the required return of investors is not influenced by the firm's dividend policy, and thus dividend policy is irrelevant.

T F 7. The payment of a stock dividend is a use of funds rather than a shifting of funds between capital accounts.

T F 8. Paying a fixed or increasing dividend increases owners' wealth because it eliminates uncertainty about the frequency and magnitude of dividends.

T F 9. Firms have no restrictions on the amount that can be paid to shareholders as dividends.

T F 10. The constant-payout ratio is the preferred method for calculating dividends.

T F 11. The clientele effect is the belief that investors see current dividends as less risky than future dividends or capital gain.

T F 12. The bird-in-the-hand argument is that different payout policies attract different types of investors but still do not change the value of the firm.

Multiple Choice

1. A dividend reinvestment plan
   a. invests dividends in the money markets.
   b. reinvests the dividends in the firm's debt.
   c. reinvests in retained earnings.
   d. reinvests in the firm's stock.

2. A firm has after-tax earnings of $2,000,000 and has declared a cash dividend of $600,000. The firm's dividend payout ratio is
   a. 70 percent.
   b. 12 percent.
   c. 30 percent.
   d. 0.70 percent.

3. The payment of cash dividends to corporate stockholders is decided by the
   a. board of directors.
   b. stockholders.
   c. top management.
   d. SEC.

4. The residual theory of dividends suggests that dividends are _________ to the value of the firm.
   a. relevant
   b. positively correlated
   c. residual
   d. irrelevant

5. Dividend policy is a form of
   a. financing policy.
   b. working capital policy.
   c. capital budgeting policy.
   d. dividend reinvestment policy.

6. Modigliani and Miller argue that when the firm has no acceptable investment opportunities, it should
   a. raise its cost of capital.
   b. keep the unneeded funds.
   c. distribute the unneeded funds to the owners.
   d. lower its cost of capital.

7. The most commonly used dividend policies are all of the following EXCEPT
   a. low-regular-and-extra.
   b. high-low.
   c. regular.
   d. constant-payout-ratio.

8. The purpose of a stock split is to
   a. lower the market price of the stock.
   b. change the firm's capital structure.
   c. raise the market price of the stock.
   d. increase dividends.

9. When common stock is repurchased and retired, the underlying motive is to
   a. reduce the stock's dividends.
   b. raise the market price of the stock.
   c. delay taxes.
   d. distribute the excess cash to the owners.

10. _________ policy is when the firm pays out a fixed dollar dividend each period.
    a. Regular dividend
    b. Constant-payout-ratio dividend
    c. Extra dividend
    d. Steady dividend

11. A _________ lowers the market price of a firm's stock by increasing the number of shares outstanding.
    a. stock dividend
    b. stock split
    c. stock repurchase
    d. dividend reinvestment policy

12. Which of the following is not a reason for a firm to repurchase stock?
    a. enhance shareholder value
    b. discourage a takeover
    c. retire shares
    d. decrease market value of the shares

13. The Board of Directors of VICEM Industries on Monday, February 4, declared a dividend payable to all holders on record on Monday, February 11, with payment on Tuesday, February 19. In order to receive a dividend from VICEM, investors would have to own the shares as of the close of trading on
    a. Monday, February 4.
    b. Wednesday, February 6.
    c. Monday, February 11.
    d. Tuesday, February 19.

14. Tyler Weinrich, a single investor in the 15 percent federal tax bracket, owns 150 shares of Newmont Exploration. The stock has risen from its purchase price of $40 to $75 dollars per share. Because the board of directors believes the Newmont Exploration will be more actively traded in the $40 to $60 range, it has just announced a 3-for-2 stock split. Assuming no other information came out about Newmont Exploration, how many shares would Tyler have and what would be their price after the stock split?
    a. 100 shares, at $40 per share
    b. 100 shares, at $50 per share
    c. 225 shares, at $50 per share
    d. 225 shares, at $60 per share

15. In early 2012 Congress passed the American Taxpayer Relief Act of 2012. As a result of the the maximum tax rate on dividends was
    a. 5 percent, or 10 percent below the maximum tax rate on capital gains.
    b. 5 percent, which is equal to the maximum tax rate on capital gains.
    c. 15 percent, which is 10 percent above the maximum tax rate on capital gains.
    d. 20 percent, which is equal to the maximum tax rate on capital gains.

16. Murray Computers has the following stockholders' equity account. The firm's common stock has a current market price of $25 per share. What is the effect on retained earnings and total stockholders' equity of a 10 percent stock dividend?

Preferred Stock	$200,000
Common Stock (15,000 shares at $2 par)	30,000
Paid-in capital in excess of par	150,000
Retained Earnings	250,000
Total stockholders' equity	$630,000

a. $246,500; $630,000
b. $250,000; $630,000
c. $212,500; $630,00
d. $212,500; $592,500

17. All of the following are significant differences between a cash dividend and a stock dividend EXCEPT:

a. Stock dividends do not affect overall stockholders' equity.
b. Stock dividends only redistribute retained earnings into common stock and additional paid-in capital accounts
c. Stock dividends decrease overall stockholders' equity.
d. Cash dividends cause a decrease in retained earnings, and hence in overall stockholders' equity.

Essay

1. What are dividend reinvestment plans and how do they work?

2. Explain why the residual theory of dividends leads Miller and Modigliani to theorize the irrelevance of dividends.

■ Chapter 13 Answer Sheet

True/False

1. T
2. F
3. F
4. T
5. T
6. T
7. F
8. T
9. F
10. F
11. F
12. F

Multiple Choice

1. D
2. C
3. A
4. D
5. A
6. C
7. B
8. A
9. D
10. A
11. B
12. D
13. B
14. C
15. D
16. C
17. C

(2.) $\frac{\$600,000}{\$2,000,000} = 30\%$

(15.)

P.S	200,000	
CS par	34,500	(1,500 shares × 3)
Paid-in	183,000	[1,500 shares × (25–3)]
RE	212,500	(250,000 – 37,500)
Total	630,000	

Essay

1. Dividend reinvestment plans allow shareholders to use the dividends they received to purchase additional shares or fractional shares of stock at little or no transaction cost. A third-party trustee can purchase the outstanding shares on the open market and then sell them to the shareholders. Another method allows the shareholders to purchase newly issued shares from the firm, usually for slightly less than the going market price.

2. According to the residual theory of dividends, dividends will be paid only when the need for additional funds is less than the earnings of the firm. The argument supporting this approach is that only sound management will ensure that the company has the money to compete effectively and this will increase share price. This tends to suggest that the required return of investors, k_s, is not influenced by the firm's dividend policy. Miller and Modigliani's argument is consistent with this and they theorize that the value of the firm is determined solely by the earning power and risk of its assets and that dividend policy has no effect on the value of the firm.

Chapter 14
Working Capital and Current Assets Management

■ Chapter Summary

In the last two chapters we studied leverage, capital structure, and dividend policy. These topics related to long-term capital. This chapter moves our discussion to the management of the firm's working capital or current assets. There is a tradeoff between the risk of having too little working capital on hand and the reduced profitability that results from having excess working capital.

In addition to discussing the issues related to determining the optimal level of working capital, this chapter also investigates the management of inventory and accounts receivable.

LG 1 *Understand working capital management, net working capital, and the related tradeoff between profitability and risk.* Working capital (or short-term financial) management focuses on managing each of the firm's current assets (inventory, accounts receivable, cash, and marketable securities) and current liabilities (accounts payable, accruals, and notes payable) in a manner that positively contributes to the firm's value. Net working capital is defined as the difference between the firm's current assets and its current liabilities, or as the portion of current assets financed with long-term funds. Firms with inadequate working capital run the risk of not being able to meet their current obligations. On the other hand, excess working capital will reduce profits.

LG 2 *Describe the cash conversion cycle, its funding requirements, and the key strategies for managing it.* The cash conversion cycle has three components: (1) average age of inventory, (2) average collection period, and (3) average payment period. The length of the cash conversion cycle determines the amount of time resources are tied up in the firm's day-to-day operations. The firm's investment in short-term assets often consists of both permanent and seasonal funding requirements. The seasonal requirements can be financed using either an aggressive (low-cost, high-risk) financing strategy or a conservative (high-cost, low-risk) financing strategy. The firm's funding decision for its cash conversion cycle ultimately depends on management's disposition toward risk and the strength of the firm's banking relationships. To minimize its reliance on negotiated liabilities, the financial manager seeks to (1) turn over inventory as quickly as possible, (2) collect accounts receivable as quickly as possible, (3) manage mail, processing, and clearing time, and (4) pay accounts payable as slowly as possible. Use of these strategies should minimize the length of the cash conversion cycle.

LG 3 *Discuss inventory management: differing views, common techniques, and international concerns.* The three types of inventory are the raw materials, work-in-process, and finished goods inventory. Different company officials will view the inventory in different ways. For example, the marketing manager will prefer a larger inventory than will the financial manager. We discuss four common inventory management methods: the ABC system, the economic order quantity (EOQ) model, the just-in-time (JIT) system, and the computerized systems for resource control—MRP, MRP II, and ERP. The models listed above provide techniques to managing inventory levels. The ABC system puts focus on those inventories that are most costly. The EOQ model can be used to determine the optimal order quantity. The JIT system is used to minimize inventory cost.

LG 4 *Explain the credit selection process and the quantitative procedure for evaluating changes in credit standards.* A firm's credit terms specify the repayment terms required of all its credit customers. The basic components include the cash discount, the cash discount period, and the credit period. The firm's collection policy is the set of procedures for collecting accounts receivable when they are due. The level of bad debt expense will be a function of both the credit standards and the collection policy. Aging of accounts receivable is a technique that indicates the proportion of the accounts receivable balance that has been outstanding for a specified period of time.

LG 5 *Review the procedures for quantitatively considering cash discounts changes, other aspects of credit terms, and credit monitoring.* Some firms that routinely give up supplier discounts may find taking these discounts could be a major source of additional income. Other firms may not be able to take the discounts because they lack alternative sources of funds. Credit monitoring, the ongoing review of accounts receivable, frequently involves use of the average collection period and an aging schedule. Firms use a number of popular collection techniques.

LG 6 *Understand the management of receipts and disbursements, including float, speeding up collections, slowing down payments, cash concentration, zero-balance accounts, and investing in marketable securities.* Float refers to funds that have been sent by the payer but are not yet usable funds to the payee. The components of float are mail time, processing time, and clearing time. Float occurs in both the average collection period and the average payment period. One technique for speeding up collections is a lockbox system. A popular technique for slowing payments is controlled disbursing.

The goal for managing operating cash is to balance the opportunity cost of nonearning balances against the transaction cost of temporary investments. Firms commonly use depository transfer checks (DTCs), ACH transfers, and wire transfers to transfer lockbox receipts to their concentration banks quickly. Zero-balance accounts (ZBAs) can be used to eliminate nonearning cash balances in corporate checking accounts. Marketable securities are short-term, interest-earning, money market instruments used by the firm to earn a return on temporarily idle funds. They may be government or nongovernment issues.

■ Chapter Notes

Net Working Capital Fundamentals

LG 1 The goal of working capital management is to manage each of the firm's current assets and current liabilities to achieve a balance between profitability and risk that maximizes the value of the owners' investment in the firm.

Current assets—Commonly called working capital, current assets represent the portion of investment that is transformed from cash to inventories to receivables and back to cash in the recurring operating cycle of the firm. Current assets are expected to be turned into cash within a year.

Current liabilities—Current liabilities represent the firm's short-term financing. They include all debts of the firm that come due in one year or less.

Net working capital—Net working capital is the difference between the firm's current assets and its current liabilities.

The Tradeoff Between Profit and Risk

Profitability is the relationship between revenues and costs generated by using the firm's assets, both current and fixed, in productive activities.

Risk is the probability that a firm will be unable to pay its bills as they come due. A firm that cannot pay its bills as they come due is said to be technically insolvent.

The effects of changes in the current assets and current liabilities on profits and risk can be demonstrated in the following chart.

Ratio	Change in Ratio	Effect on Profit	Effect on Risk
Current assets / Total assets	Increase	Decrease	Decrease
	Decrease	Increase	Increase
Current liabilities / Total assets	Increase	Increase	Increase
	Decrease	Decrease	Decrease

LG 2

The *operating cycle* of a firm is the amount of time that elapses from the point when the firm inputs material and labor into the production process to the point when cash is collected from the sale of the finished product that contains these production inputs. The operating cycle is made up of two components: (1) the average age of inventory (AAI) and (2) the average collection period of sales (ACP).

$$OC = AAI + ACP$$

The *cash conversion cycle* represents the amount of time the firm's cash is tied up between payment for production inputs and receipt of payment from the sale of the resulting finished product and can be determined by the following equation:

$$CCC = OC - APP = AAI + ACP - APP$$

where:

CCC = cash conversion cycle

OC = operating cycle

APP = average payment period (the time it takes a firm to pay for production inputs such as raw materials and labor)

Ideally, a firm would like to have a negative cash conversion cycle, but more realistically firms often have a positive cash conversion cycle, which means the firm must use nonspontaneous forms of financing to support the cash conversion cycle. When a positive cash conversion cycle is present, there are three basic strategies that a firm can employ to manage the cycle:

1. Increase inventory turnover
2. Accelerate the collection of accounts receivable
3. Stretch accounts payable, as was discussed in the previous chapter

Permanent versus Seasonal Funding Needs

The firm's financing requirements can be separated into permanent and seasonal needs. Permanent needs are the financing requirements for the firm's fixed assets plus the permanent portion of the firm's current assets. These requirements remain unchanged over the year.

The seasonal need consists of the financing requirements for the temporary current assets, which vary from time to time over the year.

Aggressive financing strategy: This strategy results in the firms' financing all of its seasonal needs and possibly some of its permanent needs with short-term funds. Financing with short-term funds is less expensive than using long-term funds. However, an increase in risk associated with the aggressive strategy results from the fact that the firm has only a limited amount of short-term borrowing capacity. If it draws too heavily on this capacity, unexpected needs for funds may become difficult to satisfy.

Conservative financing strategy: This strategy has the firm financing all projected fund requirements with long-term funds and using short-term financing only for emergencies or unexpected outflows. This strategy results in relatively lower profits because the firm uses more of the expensive long-term financing and may pay interest on unneeded funds. The conservative approach has less risk because of the high level of net working capital (i.e., liquidity) that is maintained; the firm has reserved short-term borrowing power for meeting unexpected fund demands.

Inventory Management

LG 3

The three basic types of inventory are raw materials, work-in-process, and finished goods.

Raw materials—Items purchased by the firm for use in the manufacture of a finished product

Work-in-process—Inventory that consists of all items that are currently in production
Finished goods inventory—Items that have been produced but not yet sold

Inventory should be viewed as an investment because it requires that the firm tie up its money and forego certain other earnings opportunities. In general, the higher a firm's average inventories, the larger the dollar investment and cost required; the lower its average inventories, the smaller the dollar investment and cost required.

Financial managers will tend to want to keep inventory levels low to reduce financing costs. Marketing managers will tend to want large finished goods inventories. Manufacturing managers will tend to want high raw materials and finished goods inventories. The purchasing manager may favor high raw materials inventories if quantity discounts are available for large purchases.

Techniques for Managing Inventory

Techniques that are commonly used in managing inventory are (1) The ABC system, (2) the basic economic order quantity (*EOQ*) model, (3) the reorder point, (4) the just-in-time (JIT) system, and (5) the materials requirement planning (MRP) system and enterprise resource planning (ERP) system.

The ABC system: The ABC system divides inventory into three groups. The A group includes the items that require the largest dollar investment. The B group consists of the items accounting for the next largest investment. The C group typically consists of a large number of items accounting for a relatively small dollar investment. Control of the A items should be most intensive, and control of the C group should be the least intensive.

EOQ model: This inventory management technique determines an item's optimal order quantity, which is the one that minimizes the total of its order and carrying costs.

$$EOQ = \sqrt{\frac{2 \times S \times O}{C}}$$

Where:

S = usage in units per period

O = order cost per order

C = carrying cost per unit per period

The reorder point: Once the firm has determined its economic order quantity, it must determine when to place orders. Assuming a constant usage rate for inventory, the reorder point can be determined by the following equation:

$$\text{Reorder point} = \text{lead time in days} \times \text{daily usage}$$

Because of the difficulty in precisely predicting lead times and daily usage rates, many firms typically maintain safety stocks, which are extra inventories that can be drawn down when actual outcomes are greater than expected. Note that in the absence of a reorder point, carrying costs will equal ordering costs for firms managing inventory using the *EOQ* model.

JIT system: Inventory management system minimizes inventory investment by having material inputs arrive at exactly the time they are needed for production.

With modern technology, a number of techniques have been developed for controlling inventory using computers. As a group, these systems are referred to as *computerized systems for resource control.*

MRP is a computerized system that breaks down the bill of materials for each product in order to determine what to order, when to order it, and what priorities to assign to ordering. MRP relies on EOQ and reorder point concepts to determine how much to order. This system simulates each product's bill of materials, inventory status, and manufacturing process. By monitoring the production process, order and delivery process, and inventory levels the system determines when orders should be placed for each inventory part.

MRP II is a computerized system that expands on the MRP system by incorporating other segments of the business, such as finance, accounting, marketing, engineering, and manufacturing into the model. In addition to inventory needs, the MRP II generates additional reports and coordinates the interaction of all areas impacted by inventory and production.

Enterprise resource planning (*ERP*) expands on both MRP systems by extending the information in the model to include data concerning suppliers and customers. By including both internal and external information this system is designed to eliminate production delays.

Accounts Receivable Management

LG 4

A firm's credit selection activity focuses on determining whether to extend credit to a customer and how much credit to extend. To begin the credit selection process, appropriate sources of credit information and methods of credit analysis must be developed. The five Cs of credit are the traditional focus of credit investigation.

Five Cs of Credit:

1. *Character*: The applicant's record of meeting past obligations
2. *Capacity*: The applicant's ability to repay the requested credit
3. *Capital*: The applicant's debt relative to equity
4. *Collateral*: The amount of assets the applicant has available for use in securing the credit
5. *Conditions*: The current economic and business climate, as well as any unique circumstances affecting either party to the credit transaction

Obtaining Credit Information

The major external sources of credit information are as follows:

Financial statements: The firm can analyze the applicant firm's liquidity, activity, debt, and profitability through the firm's past financial statements.

Dun & Bradstreet (*D&B*): The largest mercantile credit-reporting agency in the United States is D&B. D&B subscribers are provided with credit ratings and key estimates of overall financial strength for millions of U.S. and international firms.

Credit interchange bureaus: The National Credit Interchange System is a national network of local credit bureaus that exchange information.

Direct credit information exchanges: Often firms can obtain credit information through local, regional, or national credit associations.

Bank checking: Occasionally it will be possible to obtain credit information from the applicant's bank.

Credit Scoring

Credit scoring is a procedure resulting in a score reflecting an applicant's overall credit strength, derived as a weighted average of scores on key financial and credit characteristics.

Example

Credit Scoring of Wildwood Inc.

Financial and Credit Characteristics	Score (0 to 100)	Predetermined Weight	Weighted Score
Credit references	75	0.20	15.00
Home ownership	100	0.10	10.00
Income range	85	0.25	21.25
Payment history	80	0.25	20.00
Years at address	85	0.15	12.75
Years on job	90	0.15	13.50
		Total 1.0	Credit score 92.50

Credit Score	Action
Greater than 80	Extend credit terms
60 to 79	Extend limited credit
Less than 59	Reject application

Managing international credit: Credit management is usually more difficult for firms that operate internationally because international operations typically expose a firm to exchange rate risk and because there are dangers and delays involved with shipping goods long distances.

Changing Credit Standards

The firm's credit standards are the minimum requirements for extending credit to a customer. When evaluating proposed changes in credit standards, the three major variables that should be considered are (1) sales volume, (2) the investment in accounts receivable, and (3) bad debt expenses.

The basic changes and effects on profits expected to result from the relaxation of credit standards are tabulated as follows:

Variable	Direction of Change	Effect on Profits
Sales volume	Increase	Positive
Investment in accounts receivable	Increase	Negative
Bad debt expenses	Increase	Negative

Determining the Value of Key Variables

Profit contribution from additional sales: Fixed costs are not affected by a change in the sales level of the firm, so the additional contribution from increased sales will simply be the sales price per unit minus the variable cost per unit. The additional profit contribution per unit is multiplied by the increase in sales to determine the effect on net profits. For example, Zach's Inc. is expecting a 5 percent, or 2,000 units, increase in sales as a result of a relaxation of its credit terms. Zach's product has a selling price of \$20 and a variable cost per unit of \$12. The total additional profit contribution from the increase in sales will be \$16,000 [(\$20 – \$12) × 2,000 units].

Cost of the marginal investment in accounts receivable: The cost of the marginal investment in accounts receivable can be calculated by finding the difference between the cost of carrying receivables before and after the introduction of the relaxed credit standards. When changing the credit standards, only out-of-pocket costs are examined. Fixed costs are ignored because they are sunk and are not affected by this decision. The average investment in accounts receivable can be calculated by the following formula:

$$\text{Average investment in accounts receivable} = \frac{\text{total cost of annual sales}}{\text{turnover of accounts receivable}}$$

where:

$$\text{Turnover of accounts receivable} = \frac{360}{\text{average collection period}}$$

Cost of marginal bad debts: The cost of marginal bad debts is found by taking the difference between the level of bad debts before and after the relaxation of credit standards.

Once the values of three variables have been determined, the additional profit contribution from sales must be compared to the sum of the cost of the marginal investment in accounts receivable and the cost of marginal bad debts. If the additional profit contribution is greater than marginal cost, the credit standards should be relaxed.

LG 5

Credit terms: The supplier's credit terms state the credit period and the date the credit period ends. The supplier's credit terms are often stated on the supplier's invoice in such shorthand expressions as "3/10 net 30." The credit term would read as "3 percent discount if paid within 10 days; full payment must be made within 30 days from the beginning of the credit period."

Changing Credit Terms

A firm's credit terms specify the repayment requirements that it places on all of its credit customers. Often a type of shorthand is used. For example, credit terms may be stated as 3/10 net 30, which reads as, "3 percent discount if paid in cash within 10 days after the beginning of the credit period, otherwise full payment must be received within 30 days after the beginning of the credit period." Credit terms cover three things: (1) the cash discount (in this example 3 percent), (2) the cash discount period (10 days), and (3) the credit period (30 days).

Cash discount: When a firm increases the cash discount provided in its credit terms, the following results can be expected.

Variable	Direction of Change	Effect on Profits
Sales volume	Increase	Positive
Investments in accounts receivable due to nondiscount takers paying early	Decrease	Positive
Investment in accounts receivable due to new customers	Increase	Negative
Bad debt expense	Decrease	Positive
Profit per unit	Decrease	Negative

Cash discount period: When a firm increases the cash discount period provided in its credit terms, the following results can be expected.

Variable	Direction of Change	Effect on Profits
Sales volume	Increase	Positive
Investments in accounts receivable due to nondiscount takers paying early	Decrease	Positive
Investment in accounts receivable due to discount takers still getting cash discount but paying later	Increase	Negative
Investment in accounts receivable due to new customers	Increase	Negative
Bad debt expense	Decrease	Positive
Profit per unit	Decrease	Negative

Credit period: When a firm increases the length of the credit period provided in its credit terms, the following results can be expected.

Variable	Direction of Change	Effect on Profits
Sales volume	Increase	Positive
Investment in accounts receivable	Increase	Negative
Bad debt expenses	Increase	Negative

A firm's credit terms conform to those of its industry for competitive reasons. If their terms are less restrictive than their competitors, they will attract less creditworthy customers who may default on payments. If their credit terms are too restrictive, they will lose business to their competitors.

Collection Policy

The firm's collection policy includes the procedures used to collect accounts receivable once they are due. The effectiveness of this policy can be partly evaluated by looking at the level of bad debt expenses.

A number of collection techniques are used by the firm. The basic techniques are listed in the order typically followed in the collection process: letters, telephone calls, personal visits, collection agencies, and legal action.

Cash Management Techniques

LG 6

Financial managers have a variety of cash techniques that can provide additional savings to the firm. These techniques provide savings by taking advantage of certain imperfections in the collection and payment systems.

Float

Float refers to funds that have been dispatched by a payer but are not yet in a form that can be spent by the payee. Float has three basic components:

1. *Mail float*—The delay between the time when a payer places payment in the mail and the time when it is received by the payee

2. *Processing float*—The delay between the receipt of a check by the payee and the deposit of it in the firm's account

3. *Clearing float*—The delay between the deposit of a check by the payee and the actual availability of the funds

Speeding Up Collections

A variety of techniques are available for speeding up the collection process and thereby reducing the collection float.

Concentration banking: Concentration banking is a collection procedure in which payments are made to regionally dispersed collection centers, and then deposited in local banks for quick clearing, thus reducing collection float by shortening mail and clearing float.

Lockboxes: The lockbox system is a collection procedure in which payers send their payments to a nearby post office box that is emptied by the firm's bank several times daily. The bank deposits the payment checks in the firm's account and reduces collection float by shortening processing float as well as mail and clearing float.

Direct sends: Direct send is a collection procedure in which the payee presents payment checks directly to the banks on which they are drawn, thus reducing clearing float.

Slowing Down Disbursements

Controlled disbursing: Controlled disbursing is the strategic use of mailing points and bank accounts to lengthen mail float and clearing float, respectively.

Playing the float: Playing the float is a method of consciously anticipating the resulting float associated with the payment process and using it to keep funds in an interest-earning form for as long as possible.

Firms that aggressively manage cash disbursements will often require some type of safety device that is utilized in the event of a shortage of funds.

In an *overdraft system*, if the firm's checking account balance is insufficient to cover all checks presented against the account, the bank will automatically lend the firm enough money to cover the amount of the overdraft.

A *zero-balance account* is a checking account in which a zero balance is maintained and the firm is required to deposit funds to cover checks drawn on the account only as they are presented for payment.

Investing in Marketable Securities

Marketable securities are short-term, interest earning, money market instruments that can easily be converted into cash. To be truly marketable, a security must have two basic characteristics:

1. *Ready market*: The market for a security should have both breadth and depth to minimize the amount of time required to convert it to cash. The breadth of a market is determined by the number of buyers, and the depth of the market is determined by its ability to absorb the purchase or sale of a large dollar amount of a particular security.

2. *Safety of principal*: There should be little or no loss in the value of a marketable security over time.

Government Issues

Government issues have relatively low yields due to their low risk and the fact that the interest income on all Treasury issues and most federal agency issues is exempt from state and local taxes.

Treasury bills—U.S. Treasury obligations issued weekly on an auction basis, having varying maturities, generally under a year, and virtually no risk. The smallest denomination Treasury bill is $10,000.

Treasury notes—U.S. Treasury obligations with initial maturities of between 1 and 10 years, paying interest at a stated rate semiannually, and having virtually no risk. Treasury notes have denominations of either $1,000 or $5,000.

Federal agency issues—Low-risk securities issued by government agencies but not guaranteed by the U.S. Treasury, having generally short maturities, and offering slightly higher yields than comparable U.S. Treasury issues. These securities have denominations of $1,000 or more.

Non-Government Issues

Non-government issues of marketable securities typically have slightly higher yields than government issues with similar maturities due to the slightly higher risks associated with them and the fact that their interest income is taxable on all levels.

Negotiable certificates of deposit (CDs)—Negotiable instruments representing specific cash deposits in commercial banks, having varying maturities and yields based on size, maturity, and prevailing money market conditions.

Commercial paper—A short-term, unsecured promissory note issued by a corporation that has a very high credit standing. These securities are generally issued in multiples of $100,000.

Banker's acceptance—Short-term, low-risk marketable securities arising from bank guarantees of business transactions that are sold by banks at a discount from their maturity value.

Eurodollar deposits—Deposits of currency not native to the country in which the bank is located. The deposits are negotiable, usually pay interest at maturity, and are typically denominated in units of $1 million.

Money market mutual funds—Professionally managed portfolios of various popular marketable securities, having instant liquidity, competitive yields, and low transaction costs.

Repurchase agreement—An agreement whereby a bank or securities dealer sells a firm specific securities and agrees to repurchase them at a specific price and time.

■ Sample Problems and Solutions

Example 1. Permanent versus Seasonal Funds Requirements

Whirlybird Parachute Stuffers, Inc. has the following distribution of current and fixed assets. Divide the firm's monthly total funds requirement into a permanent and a seasonal component. What are the monthly average permanent and seasonal funds requirements?

Month	Current Assets	Fixed Assets	Total Assets
January	$2,100	$6,000	$8,100
February	2,300	6,000	8,300
March	2,400	6,000	8,400
April	2,400	6,000	8,400
May	2,600	6,000	8,600
June	5,000	6,000	11,000
July	6,000	6,000	12,000
August	6,000	6,000	12,000
September	3,000	6,000	9,000
October	2,500	6,000	8,500
November	2,000	6,000	8,000
December	1,500	6,000	7,500

Put your answer below:

Month	Total Funds Requirement	Permanent Requirement	Seasonal Requirement
January			
February			
March			
April			
May			
June			
July			
August			
September			
October			
November			
December			

Solution

To solve this problem you must determine the minimum long-term funds needed so that the firm does not have to rely on short-term funds for at least one month. The permanent requirement is the base level of total assets that remain on the books throughout the year. Total assets are always $7,500 or more. Therefore, the permanent requirement is $7,500. The seasonal is the difference between the total funds and the permanent funds. The monthly funds requirements are shown below.

Month	Total Funds Requirement	Permanent Requirement	Seasonal Requirement
January	$8,100	$7,500	$600
February	8,300	7,500	800
March	8,400	7,500	900
April	8,400	7,500	900
May	8,600	7,500	1,100
June	11,000	7,500	3,500
July	12,000	7,500	4,500
August	12,000	7,500	4,500
September	9,000	7,500	1,500
October	8,500	7,500	1,000
November	8,000	7,500	500
December	7,500	7,500	0

From this table we see that the permanent monthly average is $7,500. The monthly seasonal average is $1,650. This is found by summing up the seasonal requirement and dividing by 12 ($19,800/12). Many bank loans require that short-term loans and lines of credit be completely retired for at least one month of the year. This is to assure that the short-term funds have not become part of the long-term financing of the firm.

Example 2. Aggressive versus Conservative Financing Strategy

Whirlybird Parachute Stuffers Inc. (see Example 1) is debating whether to follow a conservative or an aggressive financing strategy. Use the answer to Example 1 to recommend a strategy, assuming that short-term funds cost 10 percent and long-term funds cost 14 percent.

Solution

Aggressive: Remember that the aggressive strategy finances the permanent with long-term funds and the seasonal with short-term funds. The cost of financing under an aggressive strategy is found by multiplying the short-term interest rate times the average seasonal borrowing and adding the cost of long-term borrowing.

$$(10\% \times \$1{,}650) + (14\% \times \$7{,}500) = \$1{,}215$$

Conservative: Remember that the conservative strategy uses long-term funds for all financing requirements. The maximum amount required is borrowed, and in months when the entire amount is not needed, it is invested on a short-term basis. The cost of a conservative strategy is found by multiplying the long-run rate by the maximum funds needed.

$$(0.14 \times \$12{,}000) = \$1{,}680.$$

Example 3. Payment Dates

If Whirlybird Parachute Stuffers, Inc. sends a bill to Harvey Field Charters dated June 25, determine, under each of the following credit terms, when the payment must be made.

a. net 30

b. net 30 EOM

c. net 45 date of invoice

Solution

a. July 25

b. July 30

c. August 9

Example 4. Cash Conversion Cycle

Cascade Bilge Pumps has an average inventory age of 80 days, an average collection period of 70 days, and an average payment period of 60 days. Compute the firm's cash cycle and cash turnover, (frequency of its cash cycle), assuming a 360-day year.

Solution

a. The cash cycle can be found by adding the average age of inventory to the average collection period and subtracting the average payment period.

$$\text{Cash cycle} = 80 \text{ days} + 70 \text{ days} - 60 \text{ days} = 90 \text{ days}$$

b. Cash turnover is found by dividing the cash cycle into the number of days in the year.

$$\text{Cash turnover} = 360 \text{ days} \div 90 \text{ days} = 4 \text{ times}$$

Example 5. Direct Sends

Tumbleweed Development has just received four checks from the sale of properties to investors. The checks, however, are drawn on a distant bank. The following table summarizes the checks.

Check From	Amount	Number of Days Float Expected
Tardy Hardy	$30,000	5 days
Bogus Bob	160,000	6 days
Fly by Night	10,000	3 days
Flaky Jake	50,000	7 days

The firm has an opportunity cost of 11 percent. They can contract with Burgess Air to have a jet fly to each city and present the checks for payment and have the funds wired directly to Tumbleweed's home office. This would result in collection in one day rather than the number of days above. The cost for Burgess Air's service is $125 per check or $400 for all four checks. Should Tumbleweed use the direct send?

Solution

To determine whether direct send is a good idea, compute the total benefit from having the funds available for investment sooner. Do this by computing the opportunity cost/day multiplied by the number of days saved.

Check From	Amount	Times Opportunity Cost/Day	Times Number of Days ←	Benefit
Tardy Hardy	$30,000	0.0003055	4 days	$ 36.67
Bogus Bob	160,000	0.0003055	5 days	244.44
Fly by Night	10,000	0.0003055	2 days	6.11
Flaky Jake	50,000	0.0003055	6 days	91.67
		Total Benefit		$378.89

This is the number of days from the earlier table, minus 1

The opportunity cost per day is the annual interest rate divided by 360 [(0.11/360)]. Since the benefit from a direct send of Bogus Bob's check is greater than $125, it should be flown. However, the total benefit is less than $400, so the rest should be collected in the usual way.

Example 6. Accounts Receivable and Cost

Heidi's Dog Frisbees sold 100,000 frisbees last year, but based on proposed changes in credit terms, the firm anticipates that sales will increase by 10 percent, to 110,000 units, in the coming year. The firm's total fixed cost is $120,000, its variable cost per unit is $1.60, and the sale price per frisbee is $4.00. The firm expects the costs and the sales price to remain unchanged in the coming year.

a. Calculate the average cost per unit under both the present and the proposed plans.

b. Calculate the additional profit contribution from sales expected to result from implementation of the proposed plan.

Solution

a. The average cost per frisbee under each plan can be found by dividing the total cost, based upon the expected sales, by the unit level of sales.

Present plan:

Total cost = $120,000 + ($1.60) × (100,000)
= $120,000 + $160,000 = $280,000

Average cost/unit = $280,000/100,000 = $2.80

Proposed plan

Total cost = $120,000 + ($1.60) × (110,000) = $296,000

Average cost/unit = $296,000/110,000 = $2.69

b. The additional profit contribution from sales expected to result from implementation of the proposal is most easily calculated by multiplying the added units of sales by their per-unit profit contribution, which can be found by subtracting the variable cost per unit from the sale price per unit.

Additional profit = (110,000 – 100,000) × ($4.00 – $1.60) = $24,000

Example 7. Accounts Receivable

Assume that under both the current and proposed plans in Example 1, Heidi's makes all sales on credit. Under the present plan, the average collection period is 36 days; under the proposed plan, it would be 72 days. The firm's required return on investment is 20 percent.

a. Calculate the firm's average investment in accounts receivable under both the present and proposed plans.

b. Determine the firm's cost of the marginal investment in accounts receivable, based on the required return of 20 percent.

c. If bad debts are unaffected by the proposal, would you recommend the proposed plan?

Solution

a. To find the average investment in accounts receivable, divide the total cost of the firm's annual sales by the turnover of accounts receivable.

Present plan

Cost of sales = 100,000 units × \$2.80/unit = \$280,000

Turnover of accounts receivable = 360/36 =10

Average investment in A/R = \$280,000/10 = \$28,000

Proposed plan

Cost of sales = 110,000 units × \$2.69/unit = \$296,000

Turnover of A/R = 360/72 days = 5

Average investment in A/R = \$296,000/5 = \$59,200

b. The cost of the marginal investment in accounts receivable is found by multiplying the marginal investment in accounts receivable by the firm's required return on investment.

Marginal investment = \$59,200 – \$28,000 = \$31,200

Cost of the marginal investment in A/R = 0.20(\$31,200) = \$6,240

c. The proposal would be highly acceptable because the additional profit contribution from sales of \$24,000 exceeds the cost of the marginal investment in accounts receivable of \$6,240.

Example 8. Bad Debt Expense

Assume that, under the present and proposed plans, Heidi's has bad debt expenses of 1 percent and 3 percent, respectively. Calculate the cost of the marginal bad debts associated with implementation of the proposed plan. Would this added information change the decision reached in Example 2?

Solution

The cost of the marginal bad debts is found by calculating the cost of bad debts under each plan and then taking their difference.

Proposed plan:	(0.03)(110,000)(\$4.00) =	\$13,200
Present plan:	(0.01)(100,000)(\$4.00) =	\$ 4,000
		\$ 9,200

Adding the cost of the marginal bad debts, $9,200, to the cost of the marginal investment in accounts receivable, $6,240, results in a total cost of the proposed plan equaling $15,440. Comparing this cost to the $24,000 additional profit contribution from marginal sales indicates that the proposed plan would still be acceptable. The marginal profit would be $8,560 [$24,000 – $15,440].

Example 9. ABC Inventory

Alena's Health Foods has only 10 different items in its inventory. The average number of these items in inventory and their associated unit costs are given below. If the firm wishes to apply an ABC inventory control system, indicate a suggested breakdown of these items into A, B, C classifications.

Item Number	Average Number of Units in Inventory	Average Cost per Unit
1	60	$1.00
2	150	9.00
3	400	0.90
4	1,000	25.00
5	3,000	4.00
6	620	10.00
7	30	200.00
8	25,000	0.02
9	100	10.00
10	20,000	0.08

Solution

There is no one correct answer to this problem. A reasonable classification can be made by first determining the average dollar investment in each item by multiplying the average number of units by their average unit cost.

Item Number	Average Dollar Investment
1	$ 60
2	1,350
3	360
4	25,000
5	12,000
6	6,200
7	6,000
8	500
9	1,000
10	1,600

From a subjective evaluation of the average dollar investments, the following classification can be made.

Class	Item Number
A	4, 5
B	6, 7
C	1, 2, 3, 8, 9, 10

The class A items are in the range of \$12,000 to \$25,000 average dollar investment; the B items are in the range of \$6,000 to \$6,200 average dollar investment; and the C items are \$1,600 or less in average dollar investment.

Example 10. EOQ

Chad's Lakeside Marina sells 100,000 gallons of fuel each year. Due to the near-perfect weather in Florida, fuel sales are constant throughout the year. Fuel can be purchased and received within 15 days. Chad's has sufficient storage capacity for up to 50,000 gallons. The firm has analyzed its inventory costs and has found that its order cost is \$250 per order and its carrying cost is \$2 per gallon per year.

a. Calculate the *EOQ* for the company's gas.

b. Calculate the total cost of the plan suggested by the *EOQ*.

c. Calculate the firm's reorder point in terms of gallons.

Solution

a. In order to calculate the *EOQ*, the values of S (100,000 gallons), O (\$250), and C (2), are substituted into the *EOQ* formula.

$$EOQ = \sqrt{\frac{2SO}{C}} = \sqrt{\frac{(2)(100,000)(\$250)}{\$2}} = 5,000 \text{ gallons}$$

b. Each component of the total inventory cost can be estimated separately. It is best to determine the order cost first because we must round off for an even number of orders.

Order cost: Dividing the annual usage by the *EOQ* results in the number of orders.

Number of orders = 100,000/5,000 gallons = 20 orders

Annual order cost = 20 × \$250 = \$5,000

Carrying cost: Multiply the average inventory by the carrying cost per unit.

Average inventory = 5,000 gallons/2 = 2,500 gallons

Annual carrying cost = 2,500 × \$2 = \$5,000

Annual inventory cost = ordering cost + carrying cost

= \$5,000 + \$5,000 = \$10,000

We know that *EOQ* and total cost were correctly calculated because annual ordering cost equals annual carrying cost.

c. The reorder point can be calculated by multiplying daily usage by the lead time, given as 15 days. Daily usage is annual usage of 100,000 gallons divided by 360 days in a year.

Daily usage = 100,000 gallons/360 days = 277.78 gals/day

Reorder point = 277.78 × 15 = 4166.67 gallons

Rounding to the next higher number, we determine that 4,167 gallons is the reorder point.

Example 11. Initiating a Cash Discount

Mac's Snow Blower Inc. currently sells 1000 snow blowers each year at $750 each with a variable cost of $400 and fixed costs of $250,000. Mac is considering initiating as cash discount of 1/10 net 30. Mac believes if he initiates the discount, he will sell 40 more snow blowers, the collection period of 30 days would go to 25 days, that 70 percent of its customers will take the 1 percent discount, and that there will be no effect on bad debt. Mac's has a cost of capital of 12 percent.

a. Calculate the additional profit contribution from sales.

b. Calculate the cost of marginal investment in A/R.

c. Calculate the cost of the discount.

d. Calculate the net profit/loss from the initiation of the proposed cash discount.

Solution

a. 40 snow blowers × ($750 – $400) = $14,000

b. Average investment presently (without discount):

($350 × 1000 units) / (365/30) = $28,767

Average investment with proposed discount:

($350 × 1040 units) / (365/25) = $24,932

Reduction in A/R: $28,767 – $24,932 = $3,835

Cost savings from reduced investment in A/R: $3,835 × 0.12 = $460

c. Cost of the discount: 1,040 units × $750 × 0.7 × 0.01 = $5,460

d. Net profit from the initiation of the proposed cash discount:$14,000 + $460 – $5,460 = $9,000

■ Study Tips

1. Short-term financial management is one of the most important and time-consuming tasks for the financial manager.

2. Recognize that there is a tradeoff between the profitability and the risk associated with various liquidity strategies.

3. Reducing inventory will cause the current ratio to fall. While we often think of high current ratios as good, effective inventory management will actually reduce it.

4. Credit management balances the risk of credit losses against losing business by being too restrictive. The goal is not to have zero credit loss. It is to maximize profit. If there are no credit losses the credit policy is probably too restrictive.

■ Student Notes

■ Sample Exam—Chapter 14

True/False

T F 1. Short-term financial management is concerned with management of the firm's current assets and current liabilities.

T F 2. The more predictable its cash inflows, the more net working capital a firm needs.

T F 3. The net working capital can be defined as the portion of the firm's current assets financed with long-term funds.

T F 4. An increase in current assets increases net working capital, thereby reducing the risk of technical insolvency.

T F 5. The aggressive financing strategy operates with minimum net working capital because only the permanent portion of the firm's current assets is being financed with long-term funds.

T F 6. The conservative strategy is less profitable than the aggressive approach because it requires the firm to pay interest on unneeded funds.

T F 7. Generally, as the firm's sales increase, accounts payable and accruals also increase.

T F 8. In credit terms, EOM (end of month) indicates that the accounts payable must be paid by the end of the month in which the merchandise has been purchased.

T F 9. If levels of cash or marketable securities are too high, the profitability of the firm will be lower than if more optimal balances were maintained.

T F 10. The firm's operating cycle is simply the sum of the average age of inventory and average payment period.

T F 11. A positive cash conversion cycle means that the firm must obtain financing to support the cash conversion cycle.

T F 12. Collection float is experienced by the payer and is a delay in the receipt of funds.

T F 13. Concentration banking is used to reduce disbursement float by lengthening the mail and clearing float components.

T F 14. The lockbox system is used to reduce collection float by shortening all three basic float components (mail, clearing, and processing).

T F 15. A zero-balance account is a checking account in which a zero balance is maintained and the bank automatically covers all checks presented against the firm's account.

T F 16. To be truly marketable, a security must have two basic characteristics: a ready market, and safety of principal.

T F 17. Federal agency issues are obligations of the U.S. Treasury and are readily accepted as low-risk securities.

T F 18. Because treasury bills are guaranteed by the U.S. Treasury, they are considered virtually risk free.

T F 19. Commercial paper is a short-term fund on deposit at commercial banks having variable yields based on size, maturity, and prevailing money market conditions.

T F 20. The yields of negotiable certificates of deposit are typically greater than those on U.S. Treasury issues and comparable to the yields on commercial paper with similar maturities.

T F 21. The cash management techniques are aimed at minimizing the firm's financing requirements by taking advantage of certain imperfections in the collection and payment system.

T F 22. Firms tend to use excess bank balances to purchase marketable securities because the rate of interest applied by banks to checking accounts is relatively low.

T F 23. The *EOQ* occurs at the point where order costs are equal to carrying costs.

T F 24. The cash discount period is the number of days after the beginning of the credit period during which the cash discount is available.

T F 25. When credit standards are relaxed it is expected that sales volume, investment in accounts receivable, and bad-debt expenses will decrease.

Multiple Choice

1. The portion of a firm's current assets financed with long-term funds may be called
   a. inventory.
   b. working capital.
   c. net working capital.
   d. accounts receivable.

2. The conversion of current assets from inventory to receivables to cash provides the cash used to pay the current liabilities, which represents a(n) ________ of cash.
   a. use, source
   b. outflow, inflow
   c. source, use
   d. inflow, outflow

3. The most difficult set of accounts to predict is
   a. current assets.
   b. fixed assets.
   c. long-term debt.
   d. current liabilities.

4. An increase in the current asset to total asset ratio has the effects of ________ on profits and ________ on risk.
   a. a decrease, an increase
   b. an increase, an increase
   c. a decrease, a decrease
   d. an increase, a decrease

For Questions 5, 6, and 7, refer to the following information.

Encore Inc.

Assets		**Liabilities and Equity**	
Current assets	$ 30,000	Current liabilities	$ 25,000
Fixed assets	75,000	Long-term debt	35,000
		Equity	45,000
Total	$105,000	Total	$105,000

The company earns 7 percent on current assets and 20 percent on fixed assets.

5. The firm's current ratio is
   a. 0.2857.
   b. 0.375.
   c. 1.2.
   d. 2.5.

6. The firm's net working capital is
   a. $10,000.
   b. –$5,000.
   c. $5,000.
   d. –$10,000.

7. The firm's annual profit on total assets is
   a. $4,100.
   b. $17,100.
   c. $18,900.
   d. $3,000.

8. In economic conditions characterized by a scarcity of short-term funds, a firm would best choose the __________ financing strategy.
   a. permanent
   b. conservative
   c. aggressive
   d. seasonal

9. One of the most common designations for the beginning of the credit period is
   a. the end of the month.
   b. the date of invoice.
   c. the transaction date.
   d. Both a and b are true

10. Cash and __________ are the firm's most liquid assets.
    a. marketable securities
    b. accounts payable
    c. inventory
    d. accruals

11. A __________ is a future debit written against a customer's checking account for an amount agreed upon by the firm to which it is payable.
    a. float
    b. wire transfer
    c. direct send
    d. preauthorized check

12. One way to improve the cash conversion cycle is to
    a. speed up collections.
    b. borrow funds.
    c. reduce inventory turnover.
    d. speed up disbursements.

13. __________ refer to the cost to the firm of forgone returns due to the failure to make short-term investments.

a. Carrying costs
b. Opportunity costs
c. Flotation costs
d. Administrative costs

14. A firm has an average age of inventory of 65 days, an average collection period of 30 days, and an average payment period of 20 days. The firm's operating cycle is

a. 95 days.
b. 115 days.
c. 50 days.
d. 85 days.

15. The __________ is the time period that elapses from the point when the firm sells a finished good on account to the point when the receivable is collected.

a. average payment period
b. average collection period
c. average age of inventory
d. cash conversion cycle

16. An increase in the average collection period will result in __________ in the operating cycle.

a. a decrease
b. no change
c. an undetermined change
d. an increase

17. A firm anticipates $1,000,000 in cash outlays during the coming year. If it costs $30 to convert marketable securities to cash and the marketable securities portfolio can earn an annual 9 percent rate of return, the economic conversion quantity of cash is

a. $666,666.
b. $18,257.
c. $25,820.
d. $32,489.

18. A decrease in the production time to manufacture a finished good will result in __________ in the cash conversion cycle.

a. no change
b. a decrease
c. an increase
d. an undetermined change

19. __________ refers to funds that have been dispatched by a payer but are not in a form that can be spent by the payee.
    a. A direct send
    b. Float
    c. The operating cycle
    d. The cash conversion cycle

20. Balances held to satisfy the safety motive are generally held as
    a. cash.
    b. accounts receivable.
    c. bonds.
    d. marketable securities.

21.. A firm with an average age of inventory of 80 days, an average collection period of 60 days, and an average payment period of 50 days has a cash conversion cycle of
    a. 1.6 times.
    b. 140 days.
    c. 2.8 times.
    d. 90 days.

22. Samantha pays her bills immediately upon receipt. Careful analysis of her payments indicates that she pays her average bill 11 days before the due date. She makes about $4,500 per month and has average monthly bills of $2,400, plus another $850 in cash payments at the grocery story, gasoline station, and similar establishments. Marketable securities are earning 5.3 percent, which is 4 percent more than the current yield on her savings interest-bearing checking account of 1.3 percent. How much money could she save by slowing down her payments?
    a. $3.83
    b. $34.72
    c. $46.01
    d. $86.25

23. Cell phone service providers offer a variety of text messaging service options. One provider's standard add-on is 300 messages for $5, or approximately 1 2/3 cents per call. Alternatively, one can pay $10 for 1,000 messages. In both cases there is a $0.20 charge per additional text message. During the past month, you sent 320 text messages. Which of the services would have been the best for you?
    a. The 300-message option, at a total cost of $4.33
    b. The 300-message option, at a total cost of $5.00
    c. The 300-message option, at a total cost of $9.00
    d. The 1,000-message option, at a total cost of $10.00

24. The Polar Corporation has daily cash receipts of $90,000. A recent analysis of its collections indicated that customers' payments were in the mail an average of four days. Once received, the payments are processed in 1½ days. After payments are deposited, it takes an average of 2½ days for these receipts to clear the banking system. How much collection float (in days) does the firm currently have?
    a. 8 days
    b. 5½ days
    c. 6½ days
    d. 5 days

25. Referring to problem 24. If the firm's opportunity cost is 11 percent, would it be economically advisable for the firm to pay an annual fee of $8,000 to reduce collection float by two days?
    a. No, because the net cost would be $1,900.
    b. Yes, because the net benefit would be $3,800.
    c. No, because the net cost would be $11,800.
    d. Yes, because the net benefit would be $11,800.

Essay

1. Discuss the tradeoff between risk and profitability that results from choosing an aggressive financing policy over a conservative one.

2. Discuss the methods available to a firm to speed up collections.

Chapter 14 Answer Sheet

True/False	Multiple Choice	
1. T	1. C	
2. F	2. C	
3. T	3. A	
4. T	4. C	$\frac{\$30,000}{\$25,000} = 1.2$
5. F	5. C	
6. T	6. C	$\$30,000 - \$25,000 = \$5,000$
7. T	7. B	$(30,000 \times 0.07) + (75,000 \times 0.2) = \$17,100$
8. F	8. B	
9. T	9. D	
10. F	10. A	
11. T	11. D	
12. F	12. A	
13. F	13. B	
14. T	14. A	
15. F	15. B	
16. T	16. D	
17. F	17. C	$ECQ = \sqrt{\frac{2 \times \$30 \times \$1,000,000}{0.09}}$; $ECQ = \$25,820$
18. T	18. B	
19. F	19. B	
20. T	20. D	
21. T	21. D	
22. T	22. B	
23. T	23. C	$4 + 1.5 + 2.5 = 8$
24. T	24. A	
25. F	25. D	$(90,000 \times 0.11 \times 2) - 8,000 = 11,800$

Essay

1. The conservative strategy requires that the firm pay interest on unused funds. It is safer because the funds are on hand when needed, but more costly. The alternative saves interest cost, but is riskier because short-term funds may not be available when they are needed for emergencies and other unexpected situations.

2. Collections can be sped up by using several different management options. Concentration banking reduces float by having banks in regions where there are many collections. Lockboxes can reduce float by having a bank pick up and process collections from a post office box. Direct sends are used to collect on large checks. A firm can also use preauthorized checks, depository transfer checks, and wire transfers.

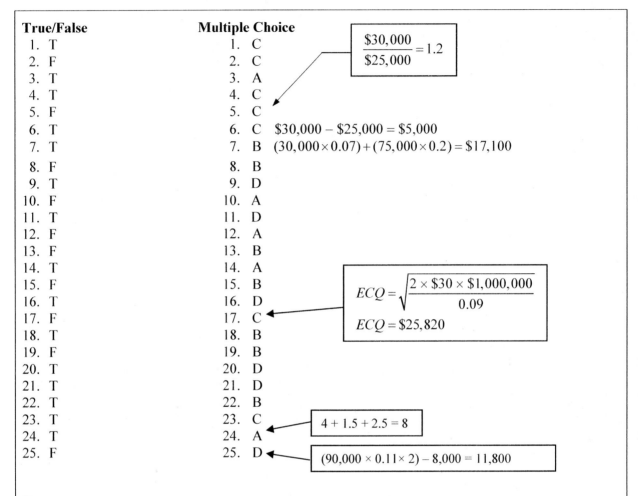

Chapter 15
Current Liabilities Management

■ Chapter Summary

In the last chapter, we studied net working capital and current asset management. We now move to the other side of the balance sheet and study current liability management. Current liabilities are short-term sources of funds and include accounts payable and short-term loans.

Some short-term financing usually is available spontaneously from increases in accounts payable and accruals that normally accompany increased sales volume. Various types of bank loans, commercial paper, and international loans are also available sources of short-term money.

LG 1 *Review accounts payable, the key components of credit terms, and the procedures for analyzing those terms.* Spontaneous liabilities arise from the normal course of doing business, the major sources being accounts payable and accruals. Therefore, they are unsecured short-term financing; that is, they are interest free and require no collateral. Credit terms may differ with respect to the credit period, cash discount, cash discount period, and beginning of the credit period. Cash discounts should be given up only when a firm in need of short-term funds must pay an interest rate on borrowing that is greater than the cost of giving up the cash discount.

LG 2 *Understand the effects of stretching accounts payable on their cost and on the use of accruals.* Stretching accounts payable can lower the cost of giving up a cash discount. Accruals, which result primarily from wage and tax obligations, are virtually interest free.

LG 3 *Describe interest rates and the basic types of unsecured bank sources of short-term loans.* Banks are the major source of unsecured short-term loans to businesses. Banks make unsecured loans in the form of single-payment notes, lines of credit, and revolving credit agreements. The interest rate on these loans is tied to the prime rate of interest by a risk premium and may be fixed or floating. It should be evaluated by using the effective annual rate. A discount loan is when the interest is paid in advance. Whether interest is paid when the loan matures or in advance affects the rate. Bank loans may take the form of a single-payment note, a line of credit, or a revolving credit agreement.

LG 4 *Discuss the basic features of commercial paper and the key aspects of international short-term loans.* Commercial paper is unsecured, short-term debt raised by firms with high credit ratings from the public and provides a very inexpensive source of funds. International sales and purchases expose firms to exchange rate risk. Such transactions are larger and of longer maturity than domestic transactions, and they can be financed by using a letter of credit, by borrowing in the local market, or through dollar-denominated loans from international banks. On transactions between subsidiaries, "netting" can be used to minimize foreign exchange fees and other transaction costs.

LG 5 *Explain the characteristics of secured short-term loans and the use of accounts receivable as short-term loan collateral.* Firms usually turn to secured sources of financing only after exhausting their unsecured options. Short-term loans are secured with collateral often in the form of accounts receivable or inventory and are more expensive than unsecured loans. Pledging is when accounts receivable is used as collateral. An alternative is factoring, which is the outright sale of accounts receivable to a financial institution.

LG 6 *Describe the various ways in which inventory can be used as short-term loan collateral.* Inventory can be used as short-term loan collateral under a floating lien, a trust receipt arrangement, or a warehouse receipt loan.

Chapter Notes

Spontaneous Sources of Short-Term Financing

LG 1 The two major spontaneous sources of short-term financing are accounts payable and accruals. They are interest free and require no collateral. Accounts payable, a major source of unsecured short-term financing, often allows the firm to take a cash discount and pay a reduced amount if payment is made by an earlier date than otherwise required. As sales increase, accounts payable increase, and therefore spontaneous financing increases.

Accounts Payable

Accounts payable result from transactions in which merchandise is purchased but no formal note is signed to show liability of purchaser and no interest is charged.

Credit terms: The supplier's credit terms state the credit period and the date the credit period begins. The supplier's credit terms are often stated on the supplier's invoice in such shorthand expressions as "3/10 net 30 EOM." The credit term would read as, "3 percent discount if paid within 10 days; full payment must be made within 30 days from the beginning of the credit period, which is the end of the month (EOM)."

The cost of giving up a cash discount can be found by using the following equation:

$$\text{Cost of giving up cash discount} = \frac{CD}{100\% - CD} \times \frac{365}{N}$$

Where:

CD = the stated cash discount in percentage terms

N = the number of days payment can be delayed by giving up the cash discount

The cost of giving up a cash discount can also be found using the approximation formula:

$$\text{Approximate cost of giving up cash discount} = \text{CD} \times \frac{365}{N}$$

As long as you can borrow for a rate less than the cost of giving up the discount, you should borrow and take the discount. For example if you can borrow from the bank at 12 percent and the cost of giving up the discount is 36 percent, then you should take the discount and borrow from the bank.

LG 2

Stretching accounts payable: Firms often stretch accounts payable by paying bills as late as possible without damaging their credit ratings. Such a strategy can reduce the cost of giving up a cash discount. For example, a firm that is given the credit terms of 2/10 net 30 EOM has a cost of giving up the cash discount of 36 percent. If the firm can stretch its accounts payable to 70 days without damaging its credit rating, the cost of giving up the cash discount would be reduced to 12 percent.

Accruals

Accruals are liabilities for services received for which payment has yet to be made. The most common accruals are wages and taxes. Taxes cannot be manipulated by the firm, but payment to employees can. The firm manipulates the payment of wages by changing the frequency of payment.

Unsecured Sources of Short-Term Loans

LG 3

Business can obtain unsecured short-term loans from two major sources—banks and commercial paper. Unlike spontaneous sources of financing, these loans are negotiated and have interest costs.

Bank loans: The interest rate on a bank loan is typically based upon the prime rate of interest and can be a fixed or a floating rate. On a fixed-rate loan, the rate of interest is determined as a set increment above the prime rate on the date of the loan and remains unvarying at that fixed rate until maturity. On a floating rate loan, the increment above the prime rate is initially established, and the rate of interest is allowed to "float," or vary, above prime as the prime rate varies until maturity.

Banks lend unsecured short-term funds in three basic ways. *Single payment notes* are short-term, one-time loans that are payable as a single amount at their maturity. A *line of credit* is an agreement between a commercial bank and a business specifying the amount of unsecured short-term borrowing the bank will make available to the firm over a given period of time. It is not a guaranteed loan. A *revolving credit agreement* is a line of credit guaranteed to the borrower by the bank for a stated time period, regardless of the scarcity of money.

If interest is paid *at maturity*, the *effective (or true) annual rate*—the actual rate of interest paid—for an assumed 1-year period is equal to

$$\frac{\text{Interest}}{\text{Amount borrowed}}$$

Most bank loans to businesses require the interest payment at maturity.

When interest is paid *in advance*, it is deducted from the loan so that the borrower actually receives less money than is requested (and less than they must repay). Loans on which interest is paid in advance are called discount loans. The *effective annual rate for a discount loan*, assuming a 1-year period, is calculated as

Loans on which interest is paid in advance by being deducted from the amount borrowed.

$$\frac{\text{Interest}}{\text{Amount borrowed} - \text{Interest}}$$

Paying interest in advance raises the effective annual rate above the stated annual rate.

With lines of credit and revolving credit agreements there may be operating-change restrictions, compensating balances, and annual cleanups. Operating restrictions are contractual restrictions a bank may impose on a firm's financial condition or operations as part of a line-of-credit agreement. A compensating balance is a required checking account balance equal to a certain percentage of the amount borrowed from a bank under a line-of-credit or revolving credit agreement. An annual cleanup is a requirement for a certain number of days during the year borrowers under a line of credit carry a zero balance. A commitment fee is a fee normally charged on a revolving credit agreement; it often applies to the average portion of the borrower's credit line.

LG 4

Commercial paper: Commercial paper is a form of financing that consists of short-term, unsecured promissory notes issued by firms with high credit standing. The interest paid by the issuer of commercial paper is determined by the size of the discount and the length of time to maturity.

Example

Whitmore Corporation has just issued $500,000 of commercial paper that has a 90-day maturity and sells for $485,000. At the end of the 90 days, the purchaser of the paper will receive $500,000 for its $485,000 investment. The interest paid on the financing is therefore 12.54 percent [($15,000 ÷ $485,000) × (365 days ÷ 90 days)].

The important difference between international and domestic transactions is that payments are often made or received in a foreign currency, which exposes companies to exchange rate risk. Typical international transactions are large in size and have long maturity dates. Therefore, companies that are involved in international trade generally have to finance larger dollar amounts for longer time periods than companies that operate domestically.

A U.S. exporter is more willing to sell goods to a foreign buyer if the transaction is covered by a letter of credit issued by a well-known bank in the buyer's home country. A letter of credit is a letter written by a company's bank to the company's foreign supplier, stating that the bank guarantees payment of an invoiced amount if all the underlying agreements are met. The letter of credit essentially substitutes the bank's reputation and creditworthiness for that of its commercial customer.

Much international trade involves transactions between corporate subsidiaries. The parent company can minimize foreign exchange fees and other transaction costs by "netting" what affiliates owe each other and paying only the net amount due, rather than having both subsidiaries pay each other the gross amounts due.

Secured Sources of Short-Term Loans

LG 5

Once a firm has exhausted its unsecured sources of short-term financing, it may be able to obtain additional short-term loans on a secured basis. Secured short-term financing has specific assets pledged as collateral and are more expensive than unsecured financing.

Lenders view secured and unsecured short-term loans as having the same degree of risk. The benefit of the collateral for a secured loan is only beneficial if the firm goes into bankruptcy. The risk associated with going bankrupt and defaulting on a loan does not change due to being secured or unsecured. However, the risk of loss is less on a secured loan.

The interest rate charged on secured short-term loans is typically higher than the interest rate on unsecured short-term loans. Typically, companies that require secured loans may not qualify for unsecured debt, and they are perceived as higher-risk borrowers by lenders. The higher rates on these secured short-term loans are attributable to the greater risk of default and the increased loan administration costs of these loans over the unsecured short-term loan.

Characteristics of Secured Short-Term Loans

Collateral: Lenders of secured short-term funds prefer collateral that has a life, or duration, that is closely matched to the term of the loan. This assures the lender that the collateral can be used to satisfy the loan in the event of default.

Terms: Under the terms of the secured loan, the lender determines the desirable percentage advance to make against the collateral. This percentage advance is the percent of the book value of the collateral that constitutes the principal of a secured loan.

The Use of Accounts Receivable As Collateral

Two commonly used means of obtaining short-term financing with accounts receivable are pledging accounts receivable and factoring accounts receivable.

Pledging accounts receivable: A pledge of accounts receivable is the use of a firm's accounts receivable as security, or collateral, to obtain a short-term loan. Accounts receivable are normally quite liquid, and they are an attractive form of short-term collateral. The lender selects acceptable accounts and lends a percentage of these accounts.

Factoring accounts receivable: Factoring accounts receivable involves the sale of the receivable, at a discount, to a factor or other financial institution. A factor is a financial institution that purchases accounts receivable from businesses. Most sales of accounts receivable to a factor are made on a non-recourse basis, meaning that the factor agrees to accept all credit risks. Commercial banks offer this type of financing.

Pledges of accounts receivable are normally made on a non-notification basis, meaning that a customer whose account has been pledged as collateral is not notified. Notification basis is when a customer whose account has been pledged (or factored) is notified to remit payment directly to the lender (or factor). This type of financing is handled by specialized financial institutions called factors; some commercial banks and commercial finance companies factor receivables.

The Use of Inventory As Collateral

LG 6

There are three ways in which a firm may use inventory as a form of collateral.

A *floating inventory lien* is a lender's claim on the borrower's general inventory as collateral for a secured loan. A *trust inventory loan* is an agreement under which the lender advances 80 percent to 100 percent of the cost of the borrower's relatively expensive inventory items in exchange for the borrower's promise to immediately repay the loan, with accrued interest, upon the sale of each item. A *warehouse receipt* is an arrangement in which the lender receives control of the pledged inventory collateral, which is warehoused by a designated agent on the lender's behalf.

■ Sample Problems and Solutions

Example 1. Loss of Loan Discounts

Determine the cost of giving up the cash discount under each of the following terms of sale.

a. 2/10 net 30

b. 3/10 net 30

c. 1/10 net 60

Solution

Solve this problem by using the equation for calculating the cost of forgoing cash discounts.

$$\text{Cost of giving up cash discount} = \frac{CD}{100\% - CD} \times \frac{365}{N}$$

Where:

CD = the stated cash discount in percent terms

N = the number of days payment can be delayed by giving up the cash discount

a. $(0.02/0.98)(365/20) = 37.24\%$

b. $(0.03/0.97)(365/20) = 56.44\%$

c. $(0.01/0.99)(365/50) = 7.37\%$

Example 2. Compensating Balance

Calculate the effective interest cost on a $300,000 line of credit with a 14 percent stated interest rate and a 15 percent compensating balance.

Solution

If the firm borrows the full amount of the loan, it actually will receive only 85 percent of the funds because 15 percent of the money must be held in a compensating balance. The annual interest in dollars on the loan would be $42,000 (i.e., 0.14 × $300,000) for a year. The effective interest cost can be found by dividing the interest paid by the amount of money the firm actually receives.

$$\text{Effective interest rate} = \$42{,}000/(0.85 \times 300{,}000) = \$42{,}000/\$255{,}000 = 16.47\%$$

Example 3. Commitment Fees

A firm is charged a 0.4 percent commitment fee on the average unused portion of a revolving credit agreement. If the agreement is for $2,000,000 and the firm's average borrowing is $1,400,000, calculate the dollar amount of the commitment fee.

Solution

Applying the 0.4 percent commitment fee to the average unused balance of $600,000 yields a commitment cost of $2,400 [i.e., 0.004 × ($2,000,000 – $1,400,000].

Example 4. Accounts Receivable as Collateral

Hole in the Wall Bank is considering making a loan secured by accounts receivable to BC and SDK Group, Inc., who wish to borrow as much as possible. The bank's policy is to accept as collateral the accounts of customers who pay within 15 days of the end of the 45-day credit period and have an average account age that is no more than 8 days beyond the customer's average payment period. The BC and SDK Group's accounts receivable balances, ages, and average payment period for each credit customer are given below. The BC and SDK Group extends net 45-day credit terms.

Customer	Accounts Receivable	Average Age of Account	Average Payment Period of Customer
A	$10,000	65	70
B	25,000	40	50
C	40,000	50	60
D	20,000	20	45
E	30,000	10	55
F	15,000	60	50
G	20,000	50	65
H	25,000	12	50

a. Calculate the dollar amount of acceptable accounts receivable collateral held by the BC and SDK Group.

b. If the bank reduces acceptable collateral by 5 percent for returns and allowances, how much acceptable collateral does the firm have?

c. If the bank will advance 70 percent against the adjusted acceptable collateral, how much can the firm borrow?

Solution

a. The acceptable accounts are those that are paid within 15 days of the end of the credit period, which would be within 60 days, and that on the average are not more than eight days beyond the customer's average payment period. For each customer the account balance, if acceptable, and its disposition are given.

Customer	Disposition	Account Balance
A	age > 60 days	0
B	OK	$25,000
C	OK	40,000
D	OK	20,000
E	OK	30,000
F	10 days beyond average pay period	0
G	Age > 60 days	0
H	OK	25,000
Total Acceptable Accounts		$140,000

b. The acceptable collateral would be 95 percent of the amount of acceptable accounts because a 5 percent reserve is maintained.

Acceptable collateral = (0.95) × ($140,000) = $133,000

c. The firm can borrow 70 percent of the acceptable collateral.

Borrowing = (0.70) × ($133,000) = $93,100.

■ Study Tips

1. Spontaneous sources of unsecured short-term financing can result from the normal business operation.

2. The decision to give up the cash discount should be the result of analysis of the cost of not taking the discount.

3. Commercial banks are the principal providers of unsecured short-term loans to business.

4. A firm may exhaust its unsecured short-term borrowing capacity. That is, beyond some level of borrowing, lenders consider a firm too risky to be given an unsecured short-term loan. The next step is to arrange for secured short-term financing.

■ Student Notes

Sample Exam—Chapter 15

True/False

T F 1. Commercial paper is short-term public debt secured by accounts receivable.

T F 2. Generally, as the firm's sales increase, accounts payable and accruals also increase.

T F 3. In credit terms, EOM (end of month) indicates that the accounts payable must be paid by the end of the month in which the merchandise has been purchased.

T F 4. If a firm anticipates stretching accounts payable, its cost of forgoing a cash discount is reduced.

T F 5. The effective interest rate for a discount loan is greater than the loan's stated interest rate.

T F 6. The interest rate on a line of credit is normally stated as a fixed rate—the prime rate—plus a percent.

T F 7. Firms are able to raise finds through the sale of commercial paper more cheaply than by borrowing from a commercial bank.

T F 8. Self-liquidating loans are mainly invested in productive assets, which provide the mechanism through which the loan is repaid.

T F 9. Secured short-term financing has specific assets pledged as collateral and appears on the balance sheet as current liabilities.

T F 10. Commercial banks and other institutions do not normally consider secured loans less risky than unsecured loans and, therefore, require higher interest rates on them.

T F 11. A line of credit is an agreement between a commercial bank and a business specifying the amount of a guaranteed, unsecured short-term borrowing the bank will make available to the firm over a given period of time.

T F 12. A revolving credit agreement is an agreement between a commercial bank and a business specifying the amount of a unsecured short-term borrowing the bank will make available to the firm over a given period of time.

Multiple Choice

1. A firm purchased goods with a purchase price of $2,500 and credit terms of 2/10 net 30. The firm paid for these goods on the fifth day after the date of sale. The firm must pay _________ for the goods.
   a. $2,450
   b. $2,475
   c. $2,500
   d. $1,250

2. The cost of forgoing a cash discount under the terms of sale 2/10 net 30 (assume a 360-day year) is
   a. 12.44 percent.
   b. 8.35 percent.
   c. 24.49 percent.
   d. 36.73 percent.

3. The __________ is the lowest rate of interest charged on business loans to the best business borrowers by the nation's leading banks.
   a. prime rate
   b. commercial paper rate
   c. treasury bill rate
   d. federal funds rate

4. Much of the commercial paper is issued by
   a. commercial finance companies.
   b. small businesses.
   c. venture capitalists.
   d. small manufacturing firms.

5. Loans on which the interest is paid in advance are often called
   a. reduced-principle loans.
   b. premium loans.
   c. called loans.
   d. discount loans.

6. Collateral typically is required for a
   a. secured short-term loan.
   b. line of credit.
   c. single payment note.
   d. called loan.

7. The interest rate charged on a secured short-term loan to a corporation is typically __________ the interest rate on an unsecured loan.
   a. unrelated to
   b. the same as
   c. lower than
   d. higher than

8. __________ involves the sale of accounts receivable.
   a. A field warehouse arrangement
   b. Factoring
   c. Pledging of accounts receivable
   d. A lien

9. A loan that is made for a specific purpose for a short time (usually less than one year) and usually has its interest rate tied to prime is called a
   a. discount loan.
   b. factor.
   c. pledge.
   d. single payment note.

10. An agreement between a commercial bank and a business specifying the amount of unsecured short-term borrowing the bank will make available to the firm over a given period of time is a
   a. discount loan.
   b. factor.
   c. line of credit.
   d. single payment note.

11. Current liabilities for an individual include all of the following except
   a. outstanding home mortgage balance.
   b. student loan payments due next June.
   c. unpaid bills from your local electric company.
   d. Visa credit card balances.

12. A local automobile dealership gives Megan the choice of a low interest rate or cash discount on the purchase of a popular sports car for $14,000. Specifically, Megan will receive a $600 cash discount if she pays 4.9 percent over the next 4 years. If she forgoes the cash discount, the dealership will offer her a car at a 4-year interest rate of 2.9 percent. In order to get the good rate, Megan will have to pay $500 up front. Ignoring time-value-of-money, which of these two choices should Megan except?
   a. the 2.9 percent loan, with total payments of $15,566
   b. the 2.9 percent loan, with total payments of $16,182
   c. the 4.9 percent loan, with total payments of $16,026
   d. the 4.9 percent loan, with total payments of $16,144

13. Austin Lynd is considering borrowing $10,000 over the next 9 months (270 days) using a variable rate loan. He will be charged 1.25 percent above the prime rate, which is now 8.5 percent. Approximately what interest expense will Austin incur if interest rates fall from this level by 0.5 percent in 60 days and then rise by 1.0 percent after 150 days for the remainder of the period?
   a. $638
   b. $688
   c. $725
   d. $858

14. __________ are the major source of unsecured short-term financing for business firms.
   a. Accruals
   b. Notes payable
   c. Accounts receivable
   d. Accounts payable

15. _________ are liabilities for services received for which payment has yet to be made. The most common accounts are taxes and wages.
    a. Accounts payable
    b. Accruals
    c. Notes payable
    d. Accounts receivable

16. Financing that arises from the normal operations of the firm is said to be
    a. accrued.
    b. payable.
    c. spontaneous.
    d. expected.

17. If you work in accounts payable and have been given the credit terms 2/10 net 40, should you take the discount and use a short-term loan from the bank at 13 percent percent or should you give up the discount? **Why**?
    a. You should take the discount and use the short-term loan at the bank because the cost of giving up the discount is 24.3 percent, which is greater than the cost of the loan from the bank.
    b. You should give up the discount and use the short-term loan at the bank because the cost of giving up the discount is 24.3 percent, which is greater than the cost of the loan from the bank.
    c. You should take the discount and use the short-term loan at the bank because the cost of giving up the discount is 24.3 percent ,which is less than the cost of the loan from the bank.
    d. You should give up the discount and use the short-term loan at the bank because the cost of giving up the discount is 24.3 percent, which is less than the cost of the loan from the bank.

18. Referring to problem 17 above , would your answer change if you could stretch your accounts payable to pay in 85 days and WHY?
    a. You should take the discount and use the short-term loan at the bank because the cost of giving up the discount is 24.3 percent, which is greater than the cost of the loan from the bank.
    b. You should give up the discount and use the short-term loan at the bank because the cost of giving up the discount is 9.7 percent, which is less than the cost of the loan from the bank.
    c. You should take the discount and use the short-term loan at the bank because the cost of giving up the discount is 9.7 percent, which is less than the cost of the loan from the bank.
    d. You should give up the discount and use the short-term loan at the bank because the cost of giving up the discount is 24.3 percent, which is less than the cost of the loan from the bank.

Essay

1. What is spontaneous short-term financing and what are the major sources of this type of financing?

2. Explain why banks do not consider secured lending to be less risky than unsecured loans.

Chapter 15 Answer Sheet

True/False

1. F
2. T
3. F
4. T
5. T
6. F
7. T
8. T
9. T
10. T
11. F
12. F

Multiple Choice

1. A $\$2,500 \times (1 - 0.02) = \$2,450$
2. C $\frac{2\%}{100\% - 2\%} \times \frac{360}{20} = 36.73\%$
3. A
4. A
5. D
6. A
7. D
8. B
9. D
10. C
11. A
12. C
13. C
14. D
15. B
16. C
17. A (Cost of giving up discount: 2× (365/30))
18. B (Cost of giving up discount: 2 × (365/75))

Essay

1. Spontaneous short-term financing is financing that arises out of the normal course of doing business. It is interest free and there is no collateral. The major sources are accounts payable and accruals.

2. Having an interest in collateral reduces the amount of loss a lender will suffer in the event of a default, but the collateral does not reduce the probability of a default. What lenders want is to be repaid on schedule. The cost of administering a secured loan is greater than the cost of administering an unsecured loan, so interest rates are usually higher.

Using a Financial Calculator

The majority of business and financial calculators offer a wide range of powerful functions. Students in the introductory finance course generally do not need to utilize all of these functions and most often find them confusing. The purpose of this appendix is to provide you with an easy and quick reference guide for some of the most commonly used financial functions.

The appendix discusses the basic functions of two of the most commonly used financial calculators, the Texas Instruments TIBAII Plus and the Hewlett Packard HP10B. After every example, the step-by-step process for the TIBAII Plus is outlined with the appropriate keystrokes, the display that will be shown on the calculator, and a description of the logic behind each step. Due to space constraints, the guide for the HP10B is not accompanied by a description section. For an explanation of the logic for the HP10B, simply refer to the step in the TIBAII Plus guide that corresponds with the step in the HP10B guide. The HP10B guide is utilized by following the steps in the left-hand column and then proceeding with the steps in the right-hand column.

TIBAII Plus

The TIBAII Plus is capable of performing four sets of functions. For the purpose of this appendix, only two sets are of importance. The functions in the first set are the ones that are written on the face of the keys. The functions in the second set are initiated by pressing the white "second function" key, which is located in the second row from the top of the keypad.

The second function key will be represented by [2nd] in this appendix.

HP10B

The HP10B performs two sets of functions. Pressing the shift key invokes the function written at the top of any key. The shift key is situated on the lower left-hand side of the calculator key pad and is yellow in color.

The shift key will be denoted as [SHIFT] in this appendix.

Calculating the present value of a lump sum amount

Example: Todd estimates that it will cost his son $55,000 to attend college in 4 years. How much should he invest today, at an annual interest rate of 12 percent (compounded monthly), to be able to send his son to college in 4 years?

TIBAII Plus

Keystrokes				Display	Description
2nd	CLRTVM			0.00	Clears the time-value-of-money worksheet
2nd	P/Y	12	ENTER	P/Y = 12.00	Sets the number of periods per year to 12
2nd	Quit			0.00	Brings the calculator to standard mode
55000	*FV*			FV = 55,000.00	Records the future cash flow of $55,000
12	I/Y			I/Y = 12.00	Records the periodic rate of interest as 12%
4	2nd		×P/Y	48.00	Calculates the number of time periods as 48
N				N = 48.00	Stores the number of time periods
CPT	PV			PV = –4,114.32	Calculates the present value of $55,000 in 4 years discounted at a monthly rate of 1%

Note: The display in the last step has a negative sign because it represents a cash outflow (investment) today.

HP10B

Keystrokes			Display	Keystrokes		Display
55000	FV		55,000.00	4	×P/YR	48
12	SHIFT	P/YR	12.00	PV		–34,114.32
12	I/YR		12.00			

Common error: When an incorrect answer is computed by the calculator, the usual mistake is a mismatch between the number of periods per year and the periodic interest rate. For example, make sure that you are not using a monthly interest rate with annual compounding.

Calculating the future value of a lump sum amount

***Example*:** If Jack invests $5,000 today in an asset earning an 8 percent rate of return (compounded annually), how much will he have after 3 years?

TIBAII Plus

Keystrokes	Display	Description
[2nd] [CLRTVM]	0.00	Clears the time-value-of-money worksheet
[2nd] [P/Y] 1 [ENTER]	P/Y = 1.00	Sets the number of periods per year to 1
[2nd] [Quit]	0.00	Brings the calculator to standard mode
5000 [+/–] [PV]	PV = –5,000.00	Records the present cash outflow of $5,000
8 [I/Y]	I/Y = 8.00	Records the periodic rate of interest as 8%
3 [N]	N = 3.00	Stores the number of time periods
[CPT] [FV]	FV = 6,298.56	Calculates the future value of $5,000 after 3 years compounded at 8%

HP10B

Keystrokes	Display	Keystrokes	Display
5000 [+/–] [PV]	–5,000.00	3 [N]	3
1 [SHIFT] [P/YR]	2.00	[FV]	6,298.56
8 [I/YR]	8.00		

Calculating the present value of an annuity

***Example*:** How much should you invest today so that starting 2 years from today, you can receive $4,000 per year for the next 4 years? Assume the discount rate is 14 percent.

TIBAII Plus

Keystrokes				Display	Description
2nd	CLRTVM			0.00	Clears the time-value-of-money worksheet
2nd	P/Y	1	ENTER	P/Y = 1.00	Sets the number of periods per year to 1
2nd	Quit			0.00	Brings the calculator to standard mode
4000	PMT			PMT = 4,000.00	Records the amount of the periodic payment
14	I/Y			I/Y = 14.00	Records the annual rate of interest as 14%
4	N			N = 4.00	Stores the number of time periods
CPT	PV			PV = −11,654.85	Calculates the present value of the annuity

HP10B

Keystrokes			Display	Keystrokes		Display
4000	PMT		4,000.00	4	N	4
11	SHIFT	P/YR	1.00	PV		−11,654.85
14	I/YR		14.00			

Calculating the present value of an annuity due

***Example*:** In this case, instead of receiving payments at the end of the year, you will receive the payments at the beginning of each year. Therefore, your first payment will be received immediately.

There are two ways to calculate the present value of an annuity due:

1. You can calculate the present value of an annuity as shown in the previous section, and multiply it by $(1 + r)$; or

2. The **TIBAII Plus** allows you to set the timing of the payment. You have to set the payment mode at "Begin" and start from the first step. This method is shown below.

Keystrokes	Display	Description
2nd CLRTVM	0.00	Clears the time-value-of-money worksheet
2nd P/Y 1 ENTER	P/Y = 1.00	Sets the number of periods per year to 1
2nd BGN	END	Shows the default setting for the payment mode
2nd SET	BGN	Sets the payment mode to the beginning of the period
2nd QUIT	0.00	Brings the calculator to the standard mode
4000 PMT	PMT = 4,000.00	Records the amount of the periodic payments
14 I/Y	I/Y = 14.00	Records the periodic rate of interest as 14%
4 N	N = 4.00	Stores the number of time periods
CPT PV	PV = –13,286.53	Calculates the present value of the annuity
2nd BGN	BGN	Invokes the payment mode
2nd SET	END	Sets the payment mode to the end of the period
2nd QUIT	0.00	Brings the calculator to the standard mode

HP10B

There are two ways in which the HP10B will calculate the value of an annuity due.

1. Calculate the present value of the annuity and multiply it by $(1 + r)$.
2. The HP10B allows you to set the timing of the payment. You have to set the payment mode at "BEGIN" and start from the first step. This method is shown below.

Keystrokes			Display	Keystrokes		Display
SHIFT	BEG/END		BEGIN	14	I/YR	14.00
4000	PMT		4,000.00	4	N	4.00
1	SHIFT	P/YR	1.00	PV		−13,286.53
				SHIFT	BEG/END	

Calculating the net present value of an annuity

***Example*:** Rob thinks if he invests $100,000 by buying property today, he can get $20,000 in rent from it for each of the next 20 years (the rent will be paid semiannually). If he wants a rate of return of 15 percent (with semiannual discounting) on his investment, what is the net present value of this project?

***Notes*:** The annual rate of return will be divided by 2; the semiannual rate of return will be 7.5 percent. The number of time periods will be multiplied by 2; $20 \times 2 = 40$. The amount of annual rent will be divided by 2; $20,000/2 = $10,000.

TIBAII Plus

Keystrokes					Display	Description
2nd	CLRWork				0.00	Clears the Cash Flow worksheet
2nd	Reset			ENTER	RST 0.00	Resets all variables to zero
CF	100000			+/−	CF_0 −100,000.00	Inputs initial cash outflow
ENTER					CF_0 = −100,000.00	Stores initial cash outflow
↓	20000			÷	CO1 = 20,000.00	Calculates periodic cash inflows
2	ENTER				CO1 = 10,000.00	Stores semiannual cash inflow amount
↓	20	×	2	ENTER	FO1 = 40.00	Stores the number of times the semiannual cash inflow occurs
NPV	15	÷	2	ENTER	I = 7.50	Stores the semiannual interest rate as 7.5%
↓	CPT				NPV = 25,944.09	Calculates the net present value of the investment

HP10B

Keystrokes			Display	Keystrokes		Display
100000	+/–	CF_j	–100,000.0	15	I/YR	15.00
20000	÷ 2	CF	10,000.00	SHIFT	NPV	25,944.09
20	× 2	P/YR	2_			
SHIFT	N_j		40.00			

Calculating the net present value of a series of uneven cash flows

Example: BlueSky Corp. is considering investing in a project with the following cash flows:

Year	Amount
0	–$2,000
1	300
2	300
3	1000
4	1100

If BlueSky has to pay an annual interest rate of 10 percent, should they invest in the project?

TIBAII Plus

Keystrokes			Display	Description
2nd	CLRWork		0.00	Clears the Cash Flow worksheet
2nd	Reset	ENTER	RST 0.00	Resets all variables to zero
CF	2000	+/–	$CF_0 - 2{,}000.00$	Inputs initial cash outflow
ENTER			$CF_0 = -2{,}000.00$	Stores initial cash outflow
↓	300	ENTER	CO1 = 300.00	Stores the first cash inflow
↓		ENTER	FO1 = 1.00	Records the first cash inflow of $300
↓	300	ENTER	CO2 = 300.00	Stores the second cash inflow of $300
↓			FO2 = 1.00	Records the second cash inflow
↓	1000	ENTER	CO3 = 1,000.00	Stores the third cash inflow
↓			FO3 = 1.00	Records that the cash inflow of $1,000 occurs once
↓	1100	ENTER	CO4 = 1,100.00	Stores the fourth cash inflow
↓			FO4 = 1	Records that the cash inflow of $1,100 occurs once
NPV	10	ENTER	I = 10.00	Stores the annual interest rate as 10%
↓	CPT		NPV = 23.29	Calculates the net present value of the Investment

HP10B

Keystrokes			Display	Keystrokes			Display
2000	+/–	CF_j	–2,000.00	1100		CF_j	1100.00
300	CF		300.00	1		SHIFT	P/YR 1.00
2	SHIFT	N_j	2.00	10		I/YR	10.0000
1000	CF_j		1000.00		SHIFT	NPV	23.29

Calculating the internal rate of return of an annuity

Example: CompuLease is planning to spend $45,000 on a computer system that will be leased to a local business. The new computer system is expected to provide after-tax cash flows of $12,000 (paid annually) per year over the next 5 years. What is the internal rate of return for the investment?

TIBAII Plus

Keystrokes			Display	Description
2nd	CLRWork		0.00	Clears the Cash Flow worksheet
2nd	Reset	ENTER	RST 0.00	Resets all variables to zero
CF	45000	+/–	CF_0 –45,000.00	Inputs initial cash outflow
↓	12000	ENTER	CO1 = 12000.00	Stores the first cash inflow
↓	5		FO1 = 5.00	Stores the number of times that cash inflow of $12000 repeats
IRR	CPT		IRR = 10.42	Calculates the annual IRR of the Investment

HP10B

Keystrokes			Display
45000	+/–	CF_j	–45,000.00
1200	CF		1200.00
5	SHIFT	N_j	5.00
SHIFT	IRR/YR		10.42

Calculating the IRR of a series of uneven cash flows

Example: Hometime Inc. has the opportunity to make an investment that requires an initial investment of \$2,500,000. The estimated cash inflows from the project for the next 4 years are shown below. What is the IRR on this investment?

TIBAII Plus

Year	Cashflow
0	–\$2,500,000
1	600,000
2	800,000
3	900,000
4	1,200,000

Keystrokes			Display	Description
2nd	CLRWork		0.00	Clears the Cash Flow worksheet
2nd	Reset	ENTER	RST 0.00	Resets all variables to zero
CF	2500000	+/–	CF_0 –2,500,000.00	Inputs initial cash outflow
ENTER			CF_0= –2,500,000.00	Stores initial cash outflow
↓	600000	ENTER	CO1 = 600,000.00	Stores the first cash inflow
↓			FO1 = 1	Records that the cash inflow of \$600,000 occurs once
↓	800000	ENTER	CO2 = 800,000.00	Stores the second cash inflow
↓			FO2 = 1.00	Records that the cash inflow of \$800,000 occurs once
↓	900000	ENTER	CO3 = 900,000.00	Stores the third cash inflow
↓			FO3 = 1	Records that the cash inflow of \$900,000 occurs once
↓	1200000	ENTER	CFO4 = 1,200,000.00	Stores the fourth cash inflow
↓			FO4 = 1	Records that the cash inflow of 1,200,000.00 occurs once
IRR	CPT		IRR = 13.30	Calculates the IRR of the project

HP10B

Keystrokes			Display	Keystrokes		Display
2500000	+/–	CF_j	–2,500,000.0	1200000	CF_j	1,200,000.00
600000	CF		600,000.00	SHIFT	IRR/YR	13.30
800000	CF_j		800,000.00			
900000	CF_j		900,000.00			

Bond valuation with interest compounded annually

Example: How much would you be willing to pay for a bond today if it pays $100 in interest annually for 15 years (starting next year) and has a principal payment of $1,000? The yield to maturity is 12 percent.

Note: This question can be interpreted as that of finding the NPV of an uneven cash flow series, with the initial cash outflow equal to zero. Hence we will follow the steps used for calculating NPV to compute the current price of the bond.

TIBAII Plus

Keystrokes			Display	Description
2nd	CLRWork		0.00	Clears the Cash Flow worksheet
2nd	Reset	ENTER	RST 0.00	Resets all variables to zero
CF	0	ENTER	CF_0 0.00	Inputs initial cash outflow
↓	100	ENTER	CO1 = 100.00	Stores the first cash inflow
↓	14		FO1 = 14.00	Stores the number of times that cash inflow of $100 repeats
↓	1100	ENTER	CO2 = 1,100.00	Stores the final cash inflow of $100 plus repayment of the $1,000 principal
NPV	12	ENTER	I = 12.00	Stores the annual discount rate as 12%
↓	CPT		NPV = 863.78	Calculates the initial price of the bond

HP10B

	Keystrokes		Display		Keystrokes		Display
0	CF$_j$		0.00	1	SHIFT	P/YR	1.00
100	CF		100.00	SHIFT	NPV		863.78
14	SHIFT	N$_j$	14.00				
1100	CF$_j$		1,100.00				

Bond valuation with interest compounded semiannually

***Note*:** Most bonds pay interest semiannually. The following example will show the conversion required to calculate the current value of such bonds.

***Example*:** If the bond described in the previous section pays interest semiannually, the calculations will be as follows:

TIBAII Plus

Keystrokes				Display	Description
2nd	CLRWork			0.00	Clears the Cash Flow worksheet
2nd	Reset		ENTER	RST 0.00	Resets all variables to zero
CF		0	ENTER	CF_0 0.00	Inputs initial cash outflow
↓		100	÷	CO1 = 100.00	Calculates the semiannual interest payment
2	ENTER			CO1 = 50.00	Stores the semiannual interest payment of $50
↓		15	×	FO1 = 15.00	Calculates the number of periods when cash inflow of $50 will occur
2	−	1	ENTER	FO1 = 29	Stores the number of interest periods
↓		1050	ENTER	CO2 = 1,050.00	Stores the final cash inflow
NPV		12	÷	I = 12.00	Calculates the semiannual discount rate
2	ENTER			I = 6.00	Stores the semiannual discount rate as 6%
↓	CPT			NPV = 862.35	Calculates the initial price of the bond

HP10B

Keystrokes						Display	Keystrokes			Display
0	CF$_j$					0.00	1050	CF$_j$		1,050.00
100	÷	2	CF$_j$			50.00	2	SHIFT	P/YR	2.00
15	×	2	−	1		1_	12	I/YR		12.00
SHIFT	N$_j$					29.00	SHIFT	NPV		862.35

Financial Dictionary

ABC system: Inventory management technique that divides inventory into three categories of descending importance based on the dollar investment in each.

ability to service debts: The ability of a firm to make the contractual payments required on a scheduled basis over the life of a debt.

accept-reject approach: The evaluation of capital expenditure proposals to determine whether they meet the firm's minimum acceptance criterion.

accounting exposure: The risk resulting from the effects of changes in foreign exchange rates on the translated value of a firm's financial statement accounts denominated in a given foreign currency.

accrual method: Recognizes revenue at the point of sale and recognizes expenses when incurred.

accruals: Liabilities for services received for which payment has yet to be made.

ACH (automated clearinghouse) credits: Deposits of payroll directly into the payees' (employees') accounts. Sacrifices disbursement float but may generate goodwill for the employer.

ACH (automated clearinghouse) debits: Preauthorized electronic withdrawals from the payer's account that are then transferred to the payee's account via a settlement among banks by the automated clearinghouse. They clear in one day, thereby reducing mail, processing, and clearing float.

acquiring company: The firm in a merger transaction that attempts to acquire another firm.

activity ratios: Used to measure the speed with which various accounts are converted into sales or cash.

adjustable-rate (or floating-rate) preferred stock (ARPS): Preferred stock whose dividend rate is tied to interest rates on specific government securities.

After-tax proceeds from sale of old asset: Found by subtracting applicable taxes from the proceeds from the sale of an old asset.

agency costs: Costs borne by stockholders to prevent or minimize agency problems and to contribute to the maximization of the owners' wealth. They include monitoring and bonding expenditures, opportunity costs, and structuring expenditures.

agency problem: The likelihood that managers may place personal goals ahead of corporate goals.

aggressive financing strategy: Strategy by which the firm finances its seasonal needs, and possibly some of its permanent needs, with short-term funds and the balance of its permanent needs with long-term funds.

aging schedule: A technique used to evaluate credit and/or collection policies that indicates the proportion of the accounts receivable balance that has been outstanding for a specified period of time.

all-current-rate method: The method by which the *functional currency*—denominated financial statements of an MNC's subsidiary—is translated into the parent company's currency.

American depository receipts (ADRs): Dollar-denominated receipts for the stock of foreign companies that are held in the vaults of banks in the companies' home countries.

American depository shares (ADSs): Securities, backed by *American depository receipts (ADRs)*, that permit U.S. investors to hold shares of non-U.S. companies and trade them in U.S. markets.

annual cleanup: The requirement that for a certain number of days during the year, borrowers under a line of credit carry a zero loan balance (i.e., owe the bank nothing).

annual percentage rate (APR): In personal finance, the effective interest rate, which must by law be clearly stated to borrowers and depositors.

annualized net present value (ANPV) approach: An approach to evaluating unequal-lived projects that converts the net present value of unequal-lived, mutually exclusive projects into an equivalent (in NPV terms) annual amount.

annuity: A stream of equal annual cash flows. These cash flows can be *inflows* of returns earned on investments or *outflows* of funds invested to earn future returns.

annuity due: An annuity for which the payments occur at the beginning of each period.

articles of partnership: The written contract used to formally establish a business partnership.

ask price: The lowest price at which a security is offered for sale.

assignment: A voluntary liquidation procedure by which a firm's creditors pass the power to liquidate the firm's assets to an adjustment bureau, a trade association, or a third party, which is designated the *assignee.*

asymmetric information: The situation in which managers of a firm have more information about operations and future prospects than do investors.

authorized shares: The number of shares of common stock that a firm's corporate charter allows without further stockholder approval.

average age of inventory: Average length of time inventory is held by the firm.

average collection period: The average amount of time needed to collect accounts receivable.

average payment period: The average amount of time needed to pay accounts payable.

average tax rate: A firm's taxes divided by its taxable income.

balance sheet: Summary statement of the firm's financial position at a given point in time.

balloon payment: At the maturity of a loan, a large lump-sum payment representing the entire loan principal if the periodic payments represent only interest.

banker's acceptances: Short-term, low-risk marketable securities arising from bank guarantees of business transactions; are sold by banks at a discount from their maturity value and provide yields slightly below those on negotiable CDs and commercial paper, but higher than those on U.S. Treasury issues.

bankruptcy: Business failure that occurs when a firm's liabilities exceed the fair market value of its assets.

Bankruptcy Reform Act of 1978: The current governing bankruptcy legislation in the United States.

bar chart: The simplest type of probability distribution; shows only a limited number of outcomes and associated probabilities for a given event.

Baumol model: A model that provides for cost-efficient transactional cash balances; assumes that the demand for cash can be predicted with certainty and determines the *economic conversion quantity (ECQ).*

bearer bonds: Bonds for which payments are made to the bearer.

best efforts basis: A public offering in which the investment banker uses his or her resources to sell the security issue without taking on the risk of underwriting and is compensated on the basis of the number of securities sold.

beta coefficient: A measure of nondiversifiable risk. An *index* of the degree of movement of an asset's return in response to a change in the *market return.*

bid price: The highest price offered to purchase a security.

bird-in-the-hand argument: The belief, in support of dividend relevance theory, that current dividend payments ("a bird in the hand") reduce investor uncertainty and ultimately result in a higher value for the firm's stock.

blue sky laws: State laws aimed at regulating the sale of securities within the state and thereby protecting investors.

board of directors: Group elected by the firm's stockholders and having ultimate authority to guide corporate affairs and make general policy.

bond: Long-term debt instrument used by business and government to raise large sums of money, generally from a diverse group of lenders.

bond indenture: A complex and lengthy legal document stating the conditions under which a bond has been issued.

bond-refunding decision: The decision facing firms, when bond interest rates drop, whether to refund (refinance) existing bonds with new bonds at the lower interest rate.

book value: The strict accounting value of an asset, calculated by subtracting its accumulated depreciation from installed cost.

book value per share: The amount per share of common stock to be received if all of the firm's assets are sold for their book value and if the proceeds remaining after payment of all liabilities (including preferred stock) are divided among the common stockholders.

book value weights: Weights that use accounting values to measure the proportion of each type of capital in the firm's financial structure; used in calculating the weighted average cost of capital.

breadth of a market: A characteristic of a ready market, determined by the number of participants (buyers) in the market.

breakeven analysis (cost-volume-profit analysis): Used (1) to determine the level of operations necessary to cover all operating costs and (2) to evaluate the profitability associated with various levels of sales.

breakeven cash inflow: The minimum level of cash inflow necessary for a project to be acceptable (i.e., NPV > $0.)

breaking point: The level of *total* new financing at which the cost of one of the financing components rises, thereby causing an upward shift in the *weighted marginal cost of capital (WMCC)*.

broker market: The *securities exchanges* on which the two sides of a transaction, the buyer and seller, are brought together to trade securities.

business risk: The risk to the firm of being unable to cover operating costs.

call feature: A feature that is included in almost all corporate bond issues that gives the issuer the opportunity to repurchase bonds prior to maturity at a stated price.

call option: An option to *purchase* a specified number of shares of a stock (typically 100) on or before some future date at a stated price.

call premium: The amount by which a bond's call price exceeds its par value.

call price: The stated price at which a bond may be repurchased, by use of a *call feature*, prior to maturity.

capital: The long-term funds of the firm; all items on the right-hand side of the firm's balance sheet, excluding current liabilities.

capital asset pricing model (CAPM): Describes the relationship between the required return, or cost of common stock equity capital, and the nondiversifiable risk of the firm as measured by the beta coefficient.

capital budgeting: The process of evaluating and selecting long-term investments that are consistent with the firm's goal of owner wealth maximization.

capital budgeting process: Consists of five distinct but interrelated steps: proposal generation, review and analysis, decision making, implementation, and follow-up.

capital expenditure: An outlay of funds by the firm that is expected to produce benefits over a period of time *greater than* one year.

capital gain: The amount by which the price at which an asset was sold exceeds the asset's initial purchase price.

capital market: A financial relationship, created by institutions and arrangements, that allows suppliers and demanders of *long-term funds* to make transactions.

capital rationing: The financial situation in which a firm has only a fixed number of dollars for allocation among competing capital expenditures.

capital structure: The mix of long-term debt and equity maintained by the firm.

capitalized lease: A *financial (capital) lease* that has the present value of all its payments included as an asset and corresponding liability on the firm's balance sheet, as required by Financial Accounting Standards Board (FASB) Standard No. 13.

carrying costs: The variable costs per unit of holding an item in inventory for a specified time period.

cash: The ready currency to which all liquid assets can be reduced.

cash bonuses: Bonus money paid to management for achieving certain performance goals.

cash budget (cash forecast): A statement of the firm's planned inflows and outflows of cash that is used to estimate its short-term cash requirements.

cash conversion cycle (CCC): Represents the amount of time the firm's cash is tied up between payment for production inputs and receipt of payment from the sale of the resulting finished product; calculated as the number of days in the firm's operating cycle minus the average payment period for inputs to production.

cash disbursements: All cash outlays by the firm during a given financial period.

cash discount: A percentage deduction from the purchase price if the buyer pays within a specified time that is shorter than the credit period.

cash discount period: The number of days after the beginning of the credit period during which the cash discount is available.

cash method: Recognizes revenues and expenses only with respect to actual inflows and outflows of cash.

cash receipts: All items from which the firm receives cash inflows during a given financial period.

change in net working capital: The difference between a change in current assets and a change in current liabilities.

Chapter 11: The portion of the *Bankruptcy Reform Act of 1978* that outlines the procedures for reorganizing a failed (or failing) firm, whether its petition is filed voluntarily or involuntarily.

Chapter 7: The portion of the *Bankruptcy Reform Act of 1978* that details the procedures to be followed in liquidating a failed firm.

clearing float: The delay between the deposit of a check by the payee and the actual availability of the funds.

Clearing House Interbank Payment System (CHIPS): The most important wire transfer service; operated by international banking consortia.

clientele effect: The theory that a firm will attract shareholders whose preferences with respect to the payment and stability of dividends correspond to the payment pattern and stability of the firm itself.

closely owned stock: All common stock of a firm owned by a small group of investors, such as a family.

coefficient of variation (CV): A measure of relative dispersion used in comparing the risk of assets with differing expected returns.

collateral: The items used by a borrower to back up a loan; any assets against which a lender has a legal claim if the borrower defaults on some provision of the loan.

collateral trust bonds: Bonds secured by stock and (or) bonds that are owned by the issuer.

collection float: The delay between the time when a payer or customer deducts a payment from its checking account ledger and the time when the payee or vendor actually receives the funds in a spendable form.

collection policy: The procedures for collecting a firm's accounts receivable when they are due.

commercial finance companies: Lending institutions that make *only* secured loans—both short term and long term—to businesses.

commercial paper: A short-term, unsecured promissory note issued by a corporation that has a very high credit standing, having a yield above that paid on U.S. Treasury issues and comparable to that available on negotiable CDs with similar maturities.

commitment fee: The fee that is normally charged on a revolving credit agreement, often based on the average unused balance of the borrower's credit line.

common stock: Collectively, units of ownership interest, or equity, in a corporation.

common stock equivalents (CSEs): All contingent securities that derive a major portion of their value from their conversion privileges or common stock characteristics.

common-size income statement: An income statement in which each item is expressed as a percentage of sales.

compensating balance: A required checking account balance equal to a certain percentage of the borrower's short-term unsecured bank loan.

competitive bidding: A method of choosing an investment banker in which the banker or group of bankers that bids the highest price for a security issue is awarded the issue.

composition: A pro rata cash settlement of creditor claims by the debtor firm; a uniform percentage of each dollar owed is paid.

compounded interest: Interest earned on a given deposit that has become part of the principal at the end of a specified period.

concentration banking: A collection procedure in which payments are made to regionally dispersed collection centers, then deposited in local banks for quick clearing. Reduces collection float by shortening mail and clearing float.

conflicting rankings: Conflicts in the ranking of a given project by NPV and IRR that result from differences in the magnitude and timing of cash flows.

congeneric merger: A merger in which one firm acquires another firm that is in the same general industry but neither in the same line of business nor a supplier or customer.

conglomerate merger: A merger combining firms in unrelated businesses.

conservative financing strategy: Strategy by which the firm finances all projected fund requirements with long-term funds and uses short-term financing only for emergencies or unexpected outflows.

consolidation: The combination of two or more firms to form a completely new corporation.

constant growth model: A widely cited dividend valuation approach that assumes that dividends will grow at a constant rate that is less than the required return.

constant growth valuation (Gordon) model: Assumes that the value of a share of stock equals the present value of all future dividends (assumed to grow at a constant rate) that it is expected to provide over an infinite time horizon.

constant-payout-ratio dividend policy: A dividend policy based on the payment of a certain percentage of earnings to owners in each dividend policy.

contingent securities: Convertibles, warrants, and stock options. Their presence affects the reporting of a firm's earnings per share (EPS).

continuous compounding: Compounding of interest an infinite number of times per year at intervals of microseconds.

continuous probability distribution: A probability distribution showing all the possible outcomes and associated probabilities for a given event.

controlled disbursing: The strategic use of mailing points and bank accounts to lengthen mail float and clearing float, respectively.

controller: The officer responsible for the firm's accounting activities, such as tax management, data processing, and cost and financial accounting.

conventional cash flow pattern: An initial outflow followed by a series of inflows.

conversion feature: A feature of so-called *convertible bonds* that allows bondholders to change each bond into a specified number of shares of common stock.

conversion feature: An option that is included as part of a bond or a preferred stock issue that allows its holder to change the security into a stated number of shares of common stock.

conversion price: The per-share price that is effectively paid for common stock as the result of conversion of a convertible security.

conversion ratio: The ratio at which a convertible security can be exchanged for common stock.

conversion (or stock) value: The value of a convertible security measured in terms of the market price of the common stock into which it can be converted.

convertible bond: A bond that can be changed into a specified number of shares of common stock.

convertible preferred stock: Preferred stock that can be changed into a specified number of shares of common stock.

corporate bond: A certificate indicating that a corporation has borrowed a certain amount of money from an institution or an individual and promises to repay it in the future under clearly defined terms.

corporate restructuring: The activities involving expansion or contraction of a firm's operations or changes in its asset or financial (ownership) structure.

correlation: A statistical measure of the relationship, if any, between series of numbers representing data of any kind.

correlation coefficient: A measure of the degree of correlation between two series.

cost of a new issue of common stock: Determined by calculating the cost of common stock after considering both the amount of underpricing and the associated flotation costs.

cost of capital: The rate of return that a firm must earn on its project investments to maintain its market value and attract needed funds.

cost of common stock equity: The rate at which investors discount the expected dividends of the firm to determine its share value.

cost of giving up a cash discount: The implied rate of interest paid to delay payment of an account payable for an additional number of days.

cost of long-term debt: The after-tax cost today of raising long-term funds through borrowing.

cost of new asset: The net outflow required to acquire a new asset.

cost of preferred stock: The annual preferred stock dividend, divided by the net proceeds from the sale of the preferred stock.

cost of retained earnings: The same as the cost of an equivalent fully subscribed issue of additional common stock, which is measured by the cost of common stock equity.

coverage ratios: Ratios that measure the firm's ability to pay certain fixed charges.

credit analysis: The evaluation of credit applicants.

credit period: The number of days until full payment of an account payable is required.

credit policy: The determination of credit selection, credit standards, and credit terms.

credit scoring: A procedure resulting in a score reflecting an applicant's overall credit strength, derived as a weighted average of scores on key financial and credit characteristics.

credit selection: The decision whether to extend credit to a customer and how much credit to extend.

credit standards: The minimum requirements for extending credit to a customer.

credit terms: Specify the repayment terms required of a firm's credit customers.

creditor control: An arrangement in which the creditor committee replaces the firm's operating management and operates the firm until all claims have been settled.

cross-sectional analysis: The comparison of different firms' financial ratios at the same point in time; involves comparing the firm's ratios to those of an industry leader or to industry averages.

cumulative preferred stock: Preferred stock for which all past (unpaid) dividends in arrears must be paid along with the current dividend before payment of dividends to common stockholders.

cumulative translation adjustment: Equity reserve account on parent company's books in which translation gains and losses are accumulated.

cumulative voting system: The system under which each share of common stock is allotted a number of votes equal to the total number of corporate directors to be elected and votes can be given to *any* director(s).

current assets: Short-term assets, expected to be converted into cash within one year or less.

current expenditure: An outlay of funds by the firm resulting in benefits received within one year.

current liabilities: Short-term liabilities, expected to be converted into cash within one year.

current rate (translation) method: Technique used by U.S.-based companies to translate their foreign-currency-denominated assets and liabilities into dollars (for consideration with the parent company's financial statements).

current ratio: A measure of liquidity calculated by dividing the firm's current assets by its current liabilities.

current yield: A measure of a bond's cash return for the year; calculated by dividing the bond's annual interest payment by its current price.

date of invoice: Indicates that the beginning of the credit period is the date on the invoice for the purchase.

date of record (dividends): The date, set by the firm's directors, on which every person whose name is recorded as a stockholder will, at a specified future time, receive a declared dividend.

date of record (rights): The last date on which the recipient of a right must be the legal owner indicated in the company's stock ledger.

dealer market: The market in which the buyer and seller are not brought together directly but instead have their orders executed by dealers that make markets in a given security.

debentures: Unsecured bonds that only creditworthy firms can issue.

debt capital: All long-term borrowing incurred by the firm.

debt ratio: Measures the proportion of total assets financed by the firm's creditors.

debt-to-equity ratio: Measures the ratio of total debt to common stock equity.

debtor in possession (DIP): The term assigned to a firm that files a reorganization petition under Chapter 11 and then develops, if feasible, a reorganization plan.

degree of financial leverage (DFL): The numerical measure of the firm's financial leverage.

degree of indebtedness: Measures amount of debt relative to other significant balance sheet amounts.

degree of operating leverage (DOL): The numerical measure of the firm's operating leverage.

degree of total leverage (DTL): The numerical measure of the firm's total leverage.

Depository Institutions Deregulation and Monetary Control Act of 1980 (DIDMCA): Signaled the beginning of the "financial services revolution" by eliminating interest-rate ceilings on all accounts and permitting certain institutions to offer new types of accounts and services.

depository transfer check (DTC): An unsigned check drawn on one of the firm's bank accounts and deposited into its account at a concentration or major disbursement bank, thereby speeding up the transfer of funds.

depreciable life: Time period over which an asset is depreciated.

depreciation: The systematic charging of a portion of the costs of fixed assets against annual revenues over time.

depth of a market: A characteristic of a ready market, determined by its ability to absorb the purchase or sale of a large dollar amount of a particular security.

dilution of ownership: Occurs when a new stock issue results in each present stockholder having a claim on a smaller part of the firm's earnings than previously.

direct lease: A lease under which a lessor owns or acquires the assets that are leased to a given lessee.

direct send: A collection procedure in which the payee presents payment checks directly to the banks on which they are drawn, thus reducing clearing float.

disbursement float: The lapse between the time when a firm deducts a payment from its checking account ledger (disburses it) and the time when funds are actually withdrawn from its account.

discount: The amount by which a bond sells at a value that is less than its par, or face, value.

discount loans: Loans on which interest is paid in advance by deducting it from the amount borrowed.

discounting cash flows: The process of finding present values; the inverse of compounding interest.

diversifiable risk: The portion of an asset's risk that is attributable to firm-specific, random causes; can be eliminated through diversification.

divestiture: The selling of some of a firm's assets for various strategic motives.

dividend irrelevance theory: A theory put forth by Miller and Modigliani that, in a perfect world, the value of a firm is unaffected by the distribution of dividends and is determined solely by the earning power and risk of its assets.

dividend payout ratio: Indicates the percentage of each dollar earned that is distributed to the owners in the form of cash; calculated by dividing the firm's cash dividend per share by its earnings per share.

dividend policy: The firm's plan of action to be followed whenever a decision concerning dividends must be made.

dividend reinvestment plans (DRPs): Plans that enable stockholders to use dividends received on the firm's stock to acquire additional full or fractional shares at little or no transaction (brokerage) cost.

dividend relevance theory: The theory, attributed to Gordon and Lintner, that stockholders prefer current dividends and that there is a direct relationship between a firm's dividend policy and its market value.

dividends: Periodic distributions of earnings to the owners of stock in a firm.

double taxation: Occurs when the already once-taxed earnings of a corporation are distributed as cash dividends to its stockholders, who are then taxed again on these dividends.

Dun & Bradstreet (D&B): The largest mercantile credit-reporting agency in the United States.

DuPont formula: Relates the firm's net profit margin and total asset turnover to its return on total assets (ROA). The ROA is the product of the net profit margin and the total asset turnover.

DuPont system of analysis: Used by management as a framework for dissecting the firm's financial statements and assessing its financial condition.

earnings per share (EPS): The amount earned during the period on each outstanding share of common stock, calculated by dividing the period's total earnings available for the firm's common stockholders by the number of shares of common stock outstanding.

EBIT-EPS approach: An approach for selecting the capital structure that maximizes earnings per share (EPS) over the expected range of earnings before interest and taxes (EBIT).

economic conversion quantity (ECQ): The cost-minimizing quantity in which to convert marketable securities to cash.

economic exposure: The risk resulting from the effects of changes in foreign exchange rates on the firm's value.

economic order quantity (EOQ) model: An inventory management technique for determining an item's optimal order quantity, which is the one that minimizes the total of its order and carrying costs.

economic value added (EVA): A popular, but static, approach to investment decisions used by many firms to determine whether an investment contributes positively to the owners' wealth.

effective interest rate: In the international context, the rate equal to the nominal rate plus (or minus) any forecast appreciation (or depreciation) of a foreign currency relative to the currency of the MNC parent.

effective (true) interest rate: The rate of interest actually paid or earned; in personal finance, commonly called the *annual percentage rate (APR).*

efficient market hypothesis: Theory describing the behavior of an assumed "perfect" market in which securities are typically in equilibrium, security prices fully reflect all public information available and react swiftly to new information, and, since stocks are fairly priced, investors need not waste time looking for mispriced securities. A market that allocates funds to their most productive uses as a result of competition among wealth-maximizing investors that determines and publicizes prices that are believed to be close to their true value. An assumed "perfect" market in which there are many small investors, each having the same information and expectations with respect to securities; there are no restrictions on investment, no taxes, and no transaction costs; and all investors are rational, they view securities similarly, and they are risk-averse, preferring higher returns and lower risk.

efficient portfolio: A portfolio that maximizes return for a given level of risk or minimizes risk for a given level of return.

end of month (EOM): Indicates that the credit period for all purchases made within a given month begins on the first day of the month immediately following.

ending cash: The sum of the firm's beginning cash and its net cash flow for the period.

equipment trust certificates: Bonds used to finance "rolling stock" such as airplanes, trucks, and railroad cars.

equity capital: The long-term funds provided by the firm's owners, the stockholders.

ethics: Standards of conduct or moral judgment.

Eurobond: A bond issued by an international borrower and sold to investors in countries with currencies other than the currency in which the bond is denominated. An international bond that is sold primarily in countries other than the country of the currency in which the issue is denominated.

Eurocurrency loan market: A large number of international banks that make long-term, floating rate, hard-currency (typically U.S. dollar-denominated) loans in the form of lines of credit to international corporate and government borrowers.

Eurocurrency market: International equivalent of the domestic money market. The portion of the Euromarket that provides short-term foreign-currency financing to subsidiaries of MNCs.

Eurodollar deposits: Deposits of currency not native to the country in which the bank is located; negotiable, usually pay interest at maturity, and are typically denominated in units of $1 million. Provide yields above nearly all other marketable securities with similar maturities.

Euro-equity market: The capital market around the world that deals in international equity issues; London has become the center of Euro-equity activity.

Euromarket: The international financial market that provides for borrowing and lending currencies outside their country of origin.

European Open Market: The transformation of the European Economic Community (EC) into a single market at year-end 1992.

ex dividend: Period beginning four *business days* prior to the date of record during which a stock will be sold without the right to receive the current dividend.

ex rights: Period beginning four *business days* prior to the date of record during which a stock will be sold without announced rights being attached to it.

excess cash balance: The (excess) amount available for investment by the firm if the period's ending cash is greater than the desired minimum cash balance; assumed to be invested in marketable securities.

excess earnings accumulation tax: The tax levied by the IRS on retained earnings above $250,000, when it has determined that the firm has accumulated an excess of earnings to allow owners to delay paying ordinary income taxes.

exchange rate risk: The danger that an unexpected change in the exchange rate between the dollar and the currency in which a project's cash flows are denominated can reduce the market value of that project's cash flow.

exchange rate risk: The risk caused by varying exchange rates between two currencies.

exercise (or option) price: The price at which holders of warrants can purchase a specified number of shares of common stock.

expectations hypothesis: Theory suggesting that the yield curve reflects investor expectations about future interest rates; an increasing inflation expectation results in an upward-sloping yield curve and a decreasing inflation expectation results in a downward-sloping yield curve.

expected return: The return that is expected to be earned each period on a given asset over an infinite time horizon.

expected value of a return: The most likely return on a given asset.

extendable notes: Debt with a short-term maturity, typically 1 to 5 years, that can be renewed for a similar period at the option of the holders.

extension: An arrangement whereby the firm's creditors receive payment in full, although not immediately.

external forecast: A sales forecast based on the relationships observed between the firm's sales and certain key external economic indicators.

external funds required ("plug" figure): Under the judgmental approach for developing a pro forma balance sheet, the amount of external financing needed to bring the statement into balance.

extra dividend: An additional dividend optionally paid by the firm if earnings are higher than normal in a given period.

factor: A financial institution that specializes in purchasing accounts receivable from businesses.

factoring accounts receivable: The outright sale of accounts receivable at a discount to a *factor* or other financial institution to obtain funds.

FASB No. 52: Statement issued by the FASB requiring U.S. multinationals first to convert the financial statement accounts of foreign subsidiaries into their *functional currency* and then to translate the accounts into the parent firm's currency using the *all-current-rate method.*

federal agency issues: Low-risk securities issued by government agencies but not guaranteed by the U.S. Treasury, having generally short maturities, and offering slightly higher yields than comparable U.S. Treasury issues.

fidelity bond: A contract under which a bonding company agrees to reimburse a firm for up to a stated amount if a specified manager's dishonest act results in a financial loss to the firm.

finance: The art and science of managing money.

financial (or capital) lease: A longer-term lease than an operating lease that is noncancelable and obligates the lessee to make payments for the use of an asset over a predefined period of time; the total payments over the term of the lease are greater than the lessor's initial cost of the leased asset.

Financial Accounting Standards Board (FASB): The accounting profession's rule-setting body, which authorizes *generally accepted accounting principles (GAAP).*

Financial Accounting Standards Board (FASB) Standard No. 52: Ruling by FASB—the policysetting body of the U.S. accounting profession—that mandates that U.S.-based companies must translate their foreign-currency-denominated assets and liabilities into dollars using the *current rate (translation) method.*

financial breakeven points: The level of EBIT necessary just to cover all fixed financial costs; the level of EBIT for which EPS = $0.

financial institution: An intermediary that channels the savings of individuals, businesses, and governments into loans or investments.

financial leverage: The magnification of risk and return introduced through the use of fixed-cost financing such as debt and preferred stock. The potential use of fixed financial costs to magnify the effects of changes in earning before interest and taxes (EBIT) on the firm's earnings per share (EPS).

financial leverage multiplier (FLM): The ratio of the firm's total assets to stockholders' equity.

financial manager: Actively manages the financial affairs of any type of business, whether financial or nonfinancial, private or public, large or small, profit-seeking or not-for-profit.

financial markets: Provide a forum in which suppliers of funds and demanders of loans and investments can transact business directly.

financial merger: A merger transaction undertaken with the goal of restructuring the acquired company to improve its cash flow and unlock its hidden value.

financial planning process: Planning that begins with long-term (strategic) financial plans that in turn guide the formulation of short-term (operating) plans and budgets.

financial risk: The risk to the firm of being unable to cover required financial obligations (interest, lease payments, preferred stock dividends).

financial services: The part of finance concerned with design and delivery of advice and financial products to individuals, business, and government.

financial supermarket: An institution at which the customer can obtain a full array of the financial services now allowed under DIDMCA.

financing flows: Cash flows that result from debt and equity-financing transactions; includes incurrence and repayment of debt, cash inflow from the sale of stock, and cash outflows to repurchase stock or pay cash dividends.

finished goods inventory: Items that have been produced but not yet sold.

five C's of credit: The five key dimensions—character, capacity, capital, collateral, and conditions—used by a firm's credit analysts in their analysis of an applicant's creditworthiness.

fixed asset turnover: Measures the efficiency with which the firm has been using its *fixed*, or earning, assets to generate sales.

fixed (or semi-fixed) relationship: The constant (or relatively constant) relationship of a currency to one of the major currencies, a combination (basket) of major currencies, or some type of international foreign exchange standard.

fixed-payment coverage ratio: Measures the firm's ability to meet all fixed-payment obligations.

fixed-rate loan: A loan whose rate of interest is determined at a set increment above the prime rate and remains unvarying at that fixed rate until maturity.

flat yield curve: A yield curve that reflects relatively similar borrowing costs for both short- and longer-term loans.

float: Funds dispatched by a payer that are not yet in a form that can be spent by the payee.

floating inventory lien: A lender's claim on the borrower's general inventory as collateral for a secured loan.

floating rate bonds: A bond where the stated interest rate is adjusted periodically within stated limits in response to changes in specified money market or capital market rates.

floating relationship: The fluctuating relationship of the values of two currencies with respect to each other.

floating-rate loan: A loan whose rate of interest is established at an increment above the prime rate and is allowed to "float," or vary, above prime as the prime rate varies until maturity.

flotation costs: The total costs of issuing and selling a security.

foreign bond: A bond issued in a host country's financial market, in the host country's currency, by a foreign borrower. An international bond that is sold primarily in the country of the currency of the issue.

foreign direct investment: The transfer of capital, managerial, and technical assets to a foreign country.

foreign direct investment (FDI): The transfer, by a multinational firm, of capital, managerial, and technical assets from its home country to a host country.

foreign exchange manager: The manager responsible for monitoring and managing the firm's exposure to loss from currency fluctuations.

foreign exchange rate: The value of two currencies with respect to each other.

forward exchange rate: The rate of exchange between two currencies at some specified future date.

friendly merger: A merger transaction endorsed by the target firm's management, approved by its stockholders, and easily consummated.

fully diluted EPS: Earnings per share (EPS) calculated under the assumption that all contingent securities are converted and exercised and are therefore common stock.

functional currency: The currency of the economic environment in which a business entity primarily generates and expends cash and in which its accounts are maintained.

future value: The value of a present amount at a future date found by applying compound interest over a specified period of time.

generally accepted accounting principles (GAAP): The practice and procedure guidelines used to prepare and maintain financial records and reports; authorized by the *Financial Accounting Standards Board (FASB)*.

golden parachutes: Provisions in the employment contracts of key executives that provide them with sizable compensation if the firm is taken over; deters hostile takeovers to the extent that the cash outflows required are large enough to make the takeover unattractive.

Gordon model: A common name for the *constant growth model* that is widely cited in dividend valuation.

government security dealer: An institution that purchases for resale various government securities and other money market instruments.

greenmail: A takeover defense under which a target firm repurchases through private negotiation a large block of stock at a premium from one or more shareholders to end a hostile takeover attempt by those shareholders.

gross profit margin: Measures the percentage of each sales dollar remaining after the firm has paid for its goods.

hedging strategies: Techniques used to offset or protect against risk; in the international context these include borrowing or lending in different currencies, undertaking contracts in the forward, futures, and/or options markets, and also swapping assets/liabilities with other parties.

historic weights: Either book or market value weights based on *actual* capital structure proportions; used in calculating the weighted average cost of capital.

holders of record: Owners of the firm's shares on the *date of record.*

holding company: A corporation that has voting control of one or more other corporations.

horizontal merger: A merger of two firms *in the same line of business.*

hostile merger: A merger transaction not supported by the target firm's management, forcing the acquiring company to try to gain control of the firm by buying shares in the marketplace.

hostile takeover: A maneuver in which an outside group, without management support, tries to gain voting control of a firm by buying its shares in the marketplace.

implied price of a warrant: The price effectively paid for each warrant attached to a bond.

incentive plans: Management compensation plans that tend to tie management compensation to share price; most popular incentive plan involves the grant of *stock options*.

income bonds: Payment of interest is required only when earnings are available.

income statement: Provides a financial summary of the firm's operating results during a specified period.

incremental cash flows: The *additional* cash flows—outflows or inflows—that are expected to result from a proposed capital expenditure.

independent projects: Projects whose cash flows are unrelated to or independent of one another; the acceptance of one *does not eliminate* the others from further consideration.

informational content: The information provided by the dividends of a firm with respect to future earnings, which causes owners to bid up (or down) the price of the firm's stock.

initial investment: The relevant cash outflow for a proposed project at time zero.

installation costs: Any added costs that are necessary to place an asset into operation.

installed cost of new asset: The cost of the asset plus its installation costs; equals the asset's depreciable value.

interest rate: The compensation paid by the borrower of funds to the lender; from the borrower's point of view, the cost of borrowing funds.

intermediate cash inflows: Cash inflows received before the termination of a project.

internal forecast: A sales forecast based on a buildup, or consensus, of forecasts through the firm's own sales channels.

internal rate of return (IRR): The discount rate that equates the present value of cash inflows with the initial investment associated with a project, thereby causing NPV = $0.

internal rate of return approach: An approach to capital rationing that involves the graphic plotting of project IRRs in descending order against the total dollar investment.

international bond: A bond that is initially sold outside the country of the borrower and often distributed in several countries.

international equity market: A vibrant equity market that emerged during the past decade to allow corporations to sell large blocks of shares in several different countries simultaneously.

intracompany netting technique: A technique used by multinational firms to minimize their cash requirements by transferring across national boundaries only the net amount of payments owed between them. Sometimes bookkeeping entries are substituted for international payment.

inventory turnover: Measures the activity, or liquidity, of a firm's inventory.

inverted yield curve: A downward-sloping yield curve that indicates generally cheaper long-term borrowing costs than short-term borrowing costs.

investment banker: A financial intermediary that purchases securities from corporate and government issuers and resells them to the general public in the primary market.

investment flows: Cash flows associated with purchase and sale of both fixed assets and equity investments in other firms.

investment opportunities schedule (IOS): A ranking of investment possibilities from best (highest returns) to worst (lowest returns). The graph that plots project IRRs in descending order against total dollar investment.

involuntary reorganization: A petition initiated by an outside party, usually a creditor, for the reorganization and payment of creditors of a failed firm.

issued shares: The number of shares of common stock that have been put into circulation; they represent the sum of outstanding shares and treasury stock.

joint venture: A partnership under which the participants have contractually agreed to contribute specified amounts of money and expertise in exchange for stated proportions of ownership and profit.

judgmental approach: A method for developing the pro forma balance sheet in which the values of certain balance sheet accounts are estimated while others are calculated, using the firm's external financing as a balancing, or "plug," figure.

junk bonds: Debt rated Ba or lower by Moody's or BB or lower by Standard & Poor's.

just-in-time (JIT) system: Inventory management system that minimizes inventory investment by having material inputs arrive at exactly the time they are needed for production.

lease-versus-purchase (or lease-versus-buy) decision: The decision facing firms needing to acquire new fixed assets; whether to lease the assets or to purchase them, using borrowed funds or available liquid resources.

leasing: The process by which a firm can obtain the use of certain fixed assets for which it must make a series of contractual, periodic, tax-deductible payments.

lessee: The receiver of the services of the assets under a lease contract.

lessor: The owner of assets that are being leased.

letter of credit: A letter written by a company's bank to the company's foreign supplier, stating that the bank will guarantee payment of an invoiced amount if all the underlying agreements are met.

letter to stockholders: Typically the first element of the annual stockholders' report and the primary communication from management to the firm's owners.

leverage: Results from the use of fixed-cost assets or funds to magnify returns to the firm's owners.

leveraged buyout (LBO): An acquisition technique involving the use of a large amount of debt to purchase a firm; an example of a *financial merger.*

leveraged lease: A lease under which the lessor acts as an equity participant, supplying only about 20 percent of the cost of the asset, while a lender supplies the balance.

leveraged recapitalization: A takeover defense in which the target firm pays a large debt-financed cash dividend, increasing the firm's financial leverage and deterring the takeover attempt.

lien: A publicly disclosed legal claim on collateral.

limited partnership: A partnership corporation: An intangible business entity created by law (often called a "legal entity").

line of credit: An agreement between a commercial bank and a business specifying the amount of unsecured short-term borrowing the bank will make available to the firm over a given period of time.

liquidation value per share: The *actual* amount per share of common stock to be received if all of the firm's assets are sold, liabilities (including preferred stock) are paid, and any remaining money is divided among the common stockholders.

liquidity: A firm's ability to satisfy its short-term obligations as they come due.

liquidity preference theory: Theory suggesting that for any given issuer, long-term interest rates tend to be higher than short-term rates due to the lower liquidity and higher responsiveness to general interest rate movements of longer-term securities; causes the yield curve to be upward-sloping.

liquidity preferences: General preferences of investors for shorter-term securities.

loan amortization: The determination of the equal annual loan payments necessary to provide a lender with a specified interest return and to repay the loan principal over a specified period.

loan amortization schedule: A schedule of equal payments to repay a loan. It shows the allocation of each loan payment to interest and principal.

lockbox system: A collection procedure in which payers send their payments to a nearby post office box that is emptied by the firm's bank several times daily; the bank deposits the payment checks in the firm's account. Reduces collection float by shortening processing float as well as mail and clearing float.

long-term (strategic) financial plans: Planned long-term financial actions and the anticipated financial impact of those actions.

long-term financing: Financing with an initial maturity of more than one year.

low-regular-and-extra dividend policy: A dividend policy based on paying a low regular dividend, supplemented by an additional dividend when earnings warrant it.

macro political risk: The subjection of *all* foreign firms to political risk (takeover) by a host country because of political change, revolution, or the adoption of new policies.

mail float: The delay between the time when a payer mails a payment and the time when the payee receives it.

maintenance clauses: Provisions within an operating lease requiring the lessor to maintain the assets and to make insurance and tax payments.

majority voting system: The system whereby in the election of the board of directors, each stockholder is entitled to one vote for each share of stock owned and he or she can vote all shares for each director.

managerial finance: Concerns the duties of the financial manager in the business firm.

marginal analysis: Economic principle that states that financial decisions should be made and actions taken only when the added benefits exceed the added costs.

marginal tax rate: The rate at which additional income is taxed.

market makers: *Securities dealers* that "make markets" by offering to buy or sell certain securities at stated prices.

market premium: The amount by which the market value exceeds the straight or conversion value of a convertible security.

market return: The return on the market portfolio of all traded securities.

market risk-return function: A graph of the discount rates associated with each level of project risk.

market segmentation theory: Theory suggesting that the market for loans is segmented on the basis of maturity and that the sources of supply and demand for loans within each segment determine its prevailing interest rate; the slope of the yield curve is determined by the general relationship between the prevailing rates in each segment.

market stabilization: The process in which an underwriting syndicate places orders to buy the security that it is attempting to sell to keep the demand for the issue, and therefore its price, at the desired level.

market value weights: Weights that use market values to measure the proportion of each type of capital in the firm's financial structure; used in calculating the weighted average cost of capital.

marketable securities: Short-term debt instruments, such as U.S. Treasury bills, commercial paper, and negotiable certificates of deposit issued by government, business, and financial institutions, respectively.

materials requirement planning (MRP) system: Inventory management system that uses EOQ concepts and a computer to compare production needs to available inventory balances and determine when orders should be placed for various items on a product's *bill of materials.*

merger: The combination of two or more firms, in which the resulting firm maintains the identity of one of the firms, usually the larger one.

micro political risk: The subjection of an individual firm, a specific industry, or companies from a particular foreign country to political risk (takeover) by the host country.

Miller-Orr model: A model that provides for cost-efficient transactional cash balances; assumes uncertain cash flows and determines an upper limit and return point for cash balances.

mixed stream: A series of cash flows exhibiting any pattern other than that of an annuity.

modified accelerated cost recovery system (MACRS): System used to determine the depreciation of assets for tax purposes.

modified DuPont formula: Relates the firm's return on total assets (ROA) to its return on equity (ROE) using the *financial leverage multiplier (FLM)*.

money market: A financial relationship created between suppliers and demanders of *short-term funds*.

money market mutual funds: Professionally managed portfolios of various popular marketable securities, having instant liquidity, competitive yields, and low transaction costs.

mortgage bonds: Bonds secured by real estate or buildings.

multinational companies (MNCs): Firms that have international assets and operations in foreign markets and draw part of their total revenue and profits from such markets.

mutually exclusive projects: Projects that compete with one another, so that the acceptance of one *eliminates* the others from further consideration.

Nasdaq: An all-electronic trading platform used to execute securities trades.

national entry control systems: Comprehensive rules, regulations, and incentives aimed at regulating inflows of *foreign direct investments* involving MNCs and at the same time extracting more benefits from their presence.

near-cash: Marketable securities, which are viewed the same as cash because of their high liquidity.

negatively correlated: Descriptive of two series that move in opposite directions.

negotiable certificates of deposit (CDs): Negotiable instruments representing specific cash deposits in commercial banks, having varying maturities and yields based on size, maturity, and prevailing money market conditions. Yields are generally above those on U.S. Treasury issues and comparable to those on commercial paper with similar maturities.

negotiated offering: A security issue for which the investment banker is merely hired rather than awarded the issue through competitive bidding.

net cash flow: The mathematical difference between the firm's cash receipts and its cash disbursements in each period.

net present value (NPV): A sophisticated capital budgeting technique; found by subtracting a project's initial investment from the present value of its cash inflows discounted at a rate equal to the firm's cost of capital.

net present value approach: An approach to capital rationing that is based on the use of present values to determine the group of projects that will maximize owners' wealth.

net present value profiles: Graphs that depict the net present value of a project for various discount rates.

net proceeds: Funds actually received from the sale of a security.

net profit margin: Measures the percentage of each sales dollar remaining after all expenses, including taxes, have been deducted.

net working capital: The difference between the firm's current assets and its current liabilities, or, alternatively, the portion of current assets financed with long-term funds; can be *positive* or *negative*.

no par value: Used to describe stock that is issued without a par value, in which case the stock may be assigned a value or placed on the firm's books at the price at which it is sold.

no-par preferred stock: Preferred stock that has a specified annual dollar dividend but no stated face value.

nominal (stated) interest rate: Contractual rate of interest charged by a lender or promised by a borrower.

nominal rate of interest: The actual rate of interest charged by the supplier of funds and paid by the demander. In the international context, the stated interest rate charged on financing when only the MNC parent's currency is involved.

Non-cash charges: Expenses deducted on the income statement that do not involve an actual outlay of cash during the period.

nonconventional cash flow pattern: A pattern in which an initial outflow is *not* followed by a series of inflows.

noncumulative preferred stock: Preferred stock for which passed (unpaid) dividends do not accumulate.

nondiversifiable risk: The relevant portion of an asset's risk attributable to market factors that affect all firms; cannot be eliminated through diversification.

non-notification basis: The basis on which a borrower, having pledged an account receivable, continues to collect the account payments without notifying the account customer.

nonparticipating preferred stock: Preferred stock whose stockholders receive only the specified dividend payments.

nonrecourse basis: The basis on which accounts receivable are sold to a factor with the understanding that the factor accept all credit risks on the purchased accounts.

nonvoting common stock: Common stock that carries no voting rights; issued when the firm wishes to raise capital through the sale of common stock but does not want to relinquish its voting control.

normal probability distribution: A symmetrical probability distribution whose shape resembles a bell-shaped curve.

normal yield curve: An upward-sloping yield curve that indicates generally cheaper short-term borrowing costs than long-term borrowing costs.

North American Free Trade Agreement (NAFTA): The treaty establishing free trade and open markets between Canada, Mexico, and the United States.

notification basis: The basis on which an account customer whose account has been pledged (or factored) is notified to remit payments directly to the lender (or factor) rather than to the borrower.

offshore centers: Certain cities or states (including London, Singapore, Bahrain, Nassau, Hong Kong, and Luxembourg) that have achieved prominence as major centers for Euromarket business.

operating breakeven point: The level of sales necessary to cover all operating costs; the point at which EBIT = $0.

operating cash flow (OCF): The cash flow a firm generates from its normal operations; calculated as *net operating profits after taxes (NPOAT)* plus depreciation.

operating cash inflows: The incremental after-tax cash inflows resulting from use of a project during its life.

operating change restrictions: Contractual restrictions that a bank may impose on a firm as part of a line of credit agreement.

operating cycle (OC): The amount of time that elapses from the point at which the firm begins to build inventory to the point at which cash is collected from sale of the resulting finished product. The recurring transition of a firm's working capital from cash to inventories to receivables and back to cash.

operating flows: Cash flows directly related to production and sale of the firm's product and services.

operating lease: A cancelable contractual arrangement whereby the lessee agrees to make periodic payments to the lessor, often for five or fewer years, for an asset's services; generally, the total payments over the term of the lease are less than the lessor's initial cost of the leased asset.

operating leverage: The potential use of fixed operating costs to magnify the effects of changes in sales on the firm's earnings before interest and taxes (EBIT).

operating profit margin: Measures the percentage of profit earned on each sales dollar before interest and taxes.

operating unit: A part of a business, such as a plant, division, product line, or subsidiary that contributes to the actual operations of the firm.

optimal capital structure: The capital structure at which the weighted average cost of capital is minimized, thereby maximizing the firm's value.

option: An instrument that provides its holder with an opportunity to purchase or sell a specified asset at a stated price on or before a set expiration date.

order costs: The fixed clerical costs of placing and receiving an inventory order.

ordinary annuity: An annuity for which the payments occur at the end of each period.

ordinary income: Income earned through the sale of a firm's goods or services.

outstanding shares: The number of shares of common stock sold to the public.

over-the-counter (OTC) market: Where smaller or unlisted securities are traded.

overdraft system: Automatic coverage by the bank of all checks presented against the firm's account, regardless of the account balance.

overhanging issue: A convertible security that cannot be forced into conversion by using the call feature.

oversubscribed issue: A security issue that is sold out.

oversubscription privilege: Provides for distribution of shares for which rights were not exercised to interested shareholders on a pro rata basis at the stated subscription price.

paid-in capital in excess of par: The amount of proceeds in excess of the par value received from the original sale of common stock.

par value: Per-share value arbitrarily assigned to an issue of common stock primarily for accounting purposes.

par-value preferred stock: Preferred stock with a stated face value that is used with the specified dividend percentage to determine the annual dollar dividend.

participating preferred stock: Preferred stock that provides for dividend payments based on certain formulas allowing preferred stockholders to participate with common stockholders in the receipt of dividends beyond a specified amount.

partnership: A business owned by two or more people and operated for profit.

payable-through draft: A draft drawn on the payer's checking account, payable to a given payee but not payable on demand; approval of the draft by the payer is required before the bank pays the draft.

payback period: The exact amount of time required for a firm to recover its initial investment in a project as calculated from cash inflows.

payment date: The actual date on which the firm will mail the dividend payment to the holders of record.

payment-in-kind (PIK) preferred stock: Preferred stock that pays dividends in additional shares of preferred stock rather than cash.

pecking order: A hierarchy of financing beginning with retained earnings followed by debt financing, and finally external equity financing.

percent-of-sales method: A method for developing the pro forma income statement that expresses the cost of goods sold, operating expenses, and interest expense as a percentage of projected sales.

percentage advance: The percent of the book value of the collateral that constitutes the principal of a secured loan.

perfectly negatively correlated: Describes two negatively correlated series that have a *correlation coefficient* of + 1.

perfectly positively correlated: Describes two positively correlated series that have a *correlation coefficient* of 1.

performance plans: Management compensation plans that compensate management on the basis of proven performance measured by EPS, growth in EPS, and other ratios of return. *Performance shares* and/or *cash bonuses* are used as compensation under these plans.

performance shares: Shares of stock given to management for meeting stated performance goals.

permanent need: Financing requirements for the firm's fixed assets plus the permanent portion of the firm's current assets; these requirements remain unchanged over the year.

perpetuity: An annuity with an infinite life, making continual annual payments.

playing the float: A method of consciously anticipating the resulting float, or delay, associated with the payment process and using it to keep funds in an interest-earning form for as long as possible.

pledge of accounts receivable: The use of a firm's accounts receivable as security, or collateral, to obtain a short-term loan.

poison pill: A takeover defense in which a firm issues securities that give their holders certain rights that become effective when a takeover is attempted; these rights make the target firm less desirable to a hostile takeover.

political risk: Risk that arises from the danger that a host government might take actions that are harmful to foreign investors or from the possibility that political turmoil in a country might endanger investments made in that country by foreign nationals. The potential discontinuity or seizure of an MNC's operations in a host country due to the host's implementation of specific rules and regulations (such as nationalization, expropriation, or confiscation).

portfolio: A collection, or group, of assets.

positively correlated: Descriptive of two series that move in the same direction.

preauthorized check (PAC): A check written by the payee against a customer's checking account for a previously agreed-upon amount. Because of prior legal authorization, the check does not require the customer's signature.

preemptive rights: Allow common stockholders to maintain their proportionate ownership in the corporation when new issues are made.

preferred stock: A special form of ownership having a fixed periodic dividend that must be paid before payment of any common stock dividends.

premium: The amount by which a bond sells at a value that is greater than its par, or face, value.

present value: The current dollar value of a future amount. The amount of money that would have to be invested today at a given interest rate over a specified period to equal the future amount.

president or chief executive officer (CEO): Corporate official responsible for managing the firm's day-to-day operations and carrying out the policies established by the board of directors.

price/earnings multiple approach: A technique to estimate the firm's share value; calculated by multiplying the firm's expected earnings per share (EPS) by the average price/earnings (P/E) ratio for the industry.

price/earnings (P/E) ratio: Reflects the amount investors are willing to pay for each dollar of the firm's earnings; the higher the P/E ratio, the greater the investor confidence in the firm.

primary EPS: Earnings per share (EPS) calculated under the assumption that all contingent securities *that derive the major portion of their value from their conversion privileges or common stock characteristics* are converted and exercised, and are therefore common stock.

primary market: Financial market in which securities are initially issued; the only market in which the issuer is directly involved in the transaction.

prime rate of interest (prime rate): The lowest rate of interest charged by the nation's leading banks on business loans to their most important and reliable business borrowers.

principal: The amount of money on which interest is paid.

private placement: The sale of a new security issue, typically debt or preferred stock, directly to an investor or group of investors.

privately owned stock: All common stock of a firm owned by a single individual.

pro forma statements: Projected, or forecast, financial statements—income statements and balance sheets.

probability distribution: A model that relates probabilities to the associated outcomes.

probability: The chance that a given outcome will occur.

proceeds from sale of old asset: The cash inflows, net of any removal or cleanup costs, resulting from the sale of an existing asset.

processing float: The delay between receipt of a check by the payee and its deposit in the firm's account.

profitability: The relationship between revenues and costs generated by using the firm's assets—both current and fixed—in productive activities.

prospectus: A portion of a security registration statement filed with the SEC that details the firm's operating and financial position; it must be made available to all potential buyers.

proxy battle: The attempt by a nonmanagement group to gain control of the management of a firm through the solicitation of a sufficient number of proxy votes.

proxy statement: A statement giving the votes of a stockholder or stockholders to another party.

public offering: The nonexclusive sale of either bonds or stock to the general public.

publicly held corporations: Corporations whose stock is traded on either an organized securities exchange or the over-the-counter exchange and/or those with more than $5 million in assets and 500 or more stockholders.

publicly owned stock: Common stock of a firm owned by a broad group of unrelated individual and/or institutional investors.

purchase options: Provisions frequently included in both operating and financial leases that allow the lessee to purchase the leased asset at maturity, typically for a prespecified price.

put option: An option to sell a given number of shares of a stock (typically 100) on or before a specified future date at a stated striking price.

putable bonds: Bonds that can be redeemed at par at the option of the holder either at specific dates or when the firm takes specific actions, such as being acquired, acquiring another, or issuing a large amount of additional debt.

pyramiding: An arrangement among holding companies wherein one holding company controls other holding companies, thereby causing an even greater magnification of earnings and losses.

quarterly compounding: Compounding of interest over four periods within the year.

quick (acid-test) ratio: A measure of liquidity calculated by dividing the firm's current assets minus inventory by current liabilities.

range: A measure of an asset's risk, which is found by subtracting the return associated with the pessimistic (worst) outcome from the return associated with optimistic (best) outcome.

ranking approach: The ranking of capital expenditure projects on the basis of some predetermined measure such as the rate of return.

ratio analysis: Involves the methods of calculating and interpreting financial ratios to assess the firm's performance and status.

ratio of exchange: The ratio of the amount *paid* per share of the target company to the per-share market price of the acquiring firm.

ratio of exchange in market price: The ratio of the market price per share of the acquiring firm *paid* to each dollar of market price per share of the target firm.

raw materials inventory: Items purchased by the firm for use in the manufacture of a finished product.

real rate of interest: The rate that creates an equilibrium between the supply of savings and the demand for investment funds in a perfect world, without inflation, where funds suppliers and demanders are indifferent to the term of loans or investments because they have no *liquidity preference*, and where all outcomes are certain.

recapitalization: The reorganization procedure under which a failed firm's debts are generally exchanged for equity or the maturities of existing debts are extended.

recaptured depreciation: The portion of the sale price that is above book value and below the initial purchase price.

recovery period: The appropriate depreciable life of a particular asset as determined by MACRS.

red herring: On a prospectus, a statement, printed in red ink, indicating the tentative nature of a security offer while it is being reviewed by the SEC.

red-line method: Unsophisticated inventory management technique in which a reorder is placed when sufficient use of inventory items from a bin exposes a red line drawn inside the bin.

regular dividend policy: A dividend policy based on the payment of a fixed-dollar dividend in each period.

relevant cash flows: The incremental after-tax cash outflow (investment) and resulting subsequent inflows associated with a proposed capital expenditure.

renewal options: Provisions especially common in operating leases that grant the lessee the option to re-lease assets at their expiration.

reorder point: The point at which to reorder inventory, expressed equationally as: lead time in days × daily usage.

repurchase agreement: An agreement whereby a bank of securities dealers sells a firm specific securities and agrees to repurchase them at a specific price and time.

required return: A specified return required each period by investors for a given level of risk. The cost of funds obtained by selling an ownership (or equity) interest; it reflects the funds supplier's level of expected return.

required total financing: Amount of funds needed by the firm if the ending cash for the period is less than the desired minimum cash balance; typically represented by notes payable.

residual theory of dividends: A theory that the dividend paid by a firm should be the amount left over after all acceptable investment opportunities have been undertaken.

restrictive covenants: Contractual clauses in long-term debt agreement that place certain operating and financial constraints on the borrower.

retained earnings: The cumulative total of all earnings, net of dividends, that have been retained and reinvested in the firm since its inception.

return: The total gain or loss experienced on behalf of the owner of an investment over a given period of time; calculated by dividing the asset's change in value plus any cash distributions during the period by its beginning-of-period investment value.

return on equity (ROE): Measures the return earned on the owners' (both preferred and common stockholders') investment in the firm.

return on total assets (ROA): Measures the overall effectiveness of management in generating profits with its available assets; also called *return on investment*.

reverse stock split: A method that is used to raise the market price of a firm's stock by exchanging a certain number of outstanding shares for one new share of stock.

revolving credit agreement: A line of credit guaranteed to the borrower by the bank for a stated time period and regardless of the scarcity of money.

risk: The chance of financial loss, or more formally, the variability of returns associated with a given asset. The chance that actual outcomes may differ from those expected. The probability that a firm will be unable to pay its bills as they come due.

risk premium: The amount by which the required discount rate for a project exceeds the risk-free rate.

risk-adjusted discount rate (RADR): The rate of return that must be earned on a given project to compensate the firm's owners adequately, thereby resulting in the maintenance or improvement of share price.

risk-averse: The attitude toward risk in which an increased return would be required for an increase in risk.

risk-free rate of interest, R_F: The required return on a risk-free asset, typically a three-month U.S. Treasury bill.

risk-free rate: The rate of return that one would earn on a virtually riskless investment such as a *U.S. Treasury bill*.

risk-indifferent: The attitude toward risk in which no change in return would be required for an increase in risk.

risk-return tradeoff: The expectation that for accepting greater risk, investors must be compensated with greater returns.

risk-seeking: The attitude toward risk in which a decreased return would be accepted for an increase in risk.

S corporation: A tax-reporting entity whose earnings are taxed not as a corporation but as the incomes of its stockholders, thus avoiding the usual *double taxation* on corporate earnings.

safety motive: A motive for holding cash or near-cash—to protect the firm against being unable to satisfy unexpected demands for cash.

safety of principal: The ease of salability of a security for close to its initial value.

safety stocks: Extra inventories that can be drawn down when actual lead times and/or usage rates are greater than expected.

sale-leaseback arrangement: A lease under which the lessee sells an asset to a prospective lessor and then leases back the same asset, making fixed periodic payments for its use.

sales forecast: The prediction of the firm's sales over a given period, based on external and/or internal data, and used as the key input to the short-term financial planning process.

scenario analysis: A technique for assessing risk that uses possible alternative conditions (scenarios) to obtain a sense of the variability in a project's return.

seasonal need: Financing requirements for temporary current assets, which vary over the year.

secondary market: Financial market in which preowned securities (those that are not new issues) are traded.

secured creditors: Creditors who have specific assets pledged as collateral and in liquidation of the failed firm receive proceeds from the sale of those assets.

secured loan: A loan that has specific assets pledged as collateral.

secured short-term financing: Short-term financing (loans) obtained by pledging specific assets as collateral.

Securities and Exchange Commission (SEC): The federal regulatory body that governs the sale and listing of securities.

securities exchanges: Organizations that provide the marketplace in which firms can raise funds through the sale of new securities and purchasers can resell securities.

security agreement: The agreement between the borrower and the lender that specifies the collateral held against a secured loan.

security market line (SML): The depiction of the *capital asset pricing model (CAPM)* as a graph that reflects the required return for each level of nondiversifiable risk (beta).

selling group: A group of brokerage firms, each of which agrees to sell a portion of a security issue and expects to make a profit on the *spread* between the price at which they buy and sell the securities.

semi-annual compounding: Compounding of interest over two periods within the year.

serial bonds: An issue of bonds of which a certain proportion matures each year.

shark repellents: Anti-takeover amendments to a corporate charter that constrain the firm's ability to transfer managerial control of the firm as a result of a merger.

shelf registration: An SEC procedure that allows firms with more than $150 million in outstanding common stock to file a "master registration statement" covering a two-year period and then, during that period, to sell securities that have already been approved under the master statement.

short-term financial management: Management of current assets and current liabilities.

short-term (operating) financial plans: Planned short-term financial actions and the anticipated financial impact of those actions.

short-term self-liquidating loan: An unsecured short-term loan in which the use to which the borrowed money is put provides the mechanism through which the loan is repaid.

signal: A financing action by management that is believed to reflect its view with respect to the firm's stock value; generally, debt financing is viewed as a *positive signal* that management believes that the stock is "undervalued," and a stock issue is viewed as a *negative signal* that management believes that the stock is "overvalued."

simulation: A statistically based behavioral approach used in capital budgeting to get a feel for risk by applying predetermined probability distributions and random numbers to estimate risky outcomes.

single-payment note: A short-term, one-time loan payable as a single amount at its maturity.

sinking-fund requirement: A restrictive provision that is often included in a bond indenture providing for the systematic retirement of bonds prior to their maturity.

small (ordinary) stock dividend: A stock dividend that represents less than 20 to 25 percent of the common stock outstanding at the time the dividend is declared.

sole proprietorship: A business owned by one person and operated for his or her own profit.

speculative motive: A motive for holding cash or near-cash—to put unneeded funds to work or to be able to quickly take advantage of unexpected opportunities that may arise.

spin-off: A form of divestiture in which an operating unit becomes an independent company by issuing shares in it on a pro rata basis to the parent company's shareholders.

sponsored ADR: An ADR for which the issuing (foreign) company absorbs the legal and financial costs of creating and trading a security.

spontaneous financing: Financing that arises from the normal operations of the firm, the two major short-term sources of which are accounts payable and accruals.

spot exchange rate: The rate of exchange between two currencies on any given day.

spread: The difference between the price paid for a security by the investment banker and the sale price.

staggered funding: A way to play the float by depositing a certain proportion of a payroll or payment into the firm's checking account on several successive days *following* the actual issuance of checks.

stakeholder: Groups such as employees, customers, suppliers, creditors, and others who have a direct economic link to the firm.

standard debt provisions: Provisions in long-term debt agreements specifying certain criteria of satisfactory record keeping and reporting, tax payment, and general business maintenance on the part of the borrowing firm; normally, they do not place a burden on the financially sound business.

standard deviation: The most common statistical indicator of an asset's risk; it measures the dispersion around the *expected* value.

standby arrangement: A formal guarantee that any shares that are not subscribed or sold publicly will be purchased by the investment banker.

statement of cash flows: Provides a summary of the firm's operating, investment, and financing cash flows and reconciles them with changes in its cash and marketable securities during the period of concern.

statement of retained earnings: Reconciles the net income earned during a given year and any cash dividends paid with the change in retained earnings between the start and end of that year.

stock dividend: The payment to existing owners of a dividend in the form of stock.

stock options: An incentive allowing management to purchase stock at the market price set at the time of the grant. Options, generally extended to management, that permit purchase of the firm's common stock at a specified price over a stated period of time.

stock repurchase: The repurchasing by the firm of outstanding shares of its common stock in the marketplace; desired effects of stock repurchases are that they enhance shareholder value and/or help to discourage unfriendly takeovers.

stock rights: Provide stockholders with the privilege to purchase additional shares of stock in direct proportion to their number of owned shares.

stock split: A method that is commonly used to lower the market price of a firm's stock by increasing the number of shares belonging to each shareholder.

stock swap transaction: An acquisition method in which the acquiring firm exchanges its shares for shares of the target company according to a predetermined ratio.

stock-purchase plans: An employee fringe benefit that allows the purchase of a firm's stock at a discount or on a matching basis with a part of the cost absorbed by the firm.

stock-purchase warrant: An instrument that gives its holder the right to purchase a certain number of shares of common stock at a specified price over a certain period of time.

stockholders: The true owners of the firm by virtue of their equity in the form of preferred and common stock.

stockholders' report: Annual report required of publicly held corporations that summarizes and documents for stockholders the firm's financial activities during the past year.

straight bond value: The price at which a convertible bond would sell in the market without the conversion feature.

straight bond: A bond that is nonconvertible, having no conversion feature.

straight preferred stock: Preferred stock that is nonconvertible, having no conversion feature.

strategic merger: A merger transaction undertaken to achieve economies of scale.

stretching accounts payable: Paying bills as late as possible without damaging one's credit rating.

striking price: The price at which the holder of a call option can buy (or the holder of a put option can sell) a specified amount of stock at any time before the option's expiration date.

subordinated debentures: Claims of these bonds are not satisfied until those of the creditors holding certain (senior) debts have been fully satisfied.

subordination: In a long-term debt agreement, the stipulation that all subsequent or less important creditors agree to wait until all claims of the senior debt are satisfied before having their claims satisfied.

subscription price: The price at which stock rights are exercisable for a specified period of time; is set below the prevailing market price.

subsidiaries: The companies controlled by a holding company.

supervoting shares: Stock that carries with it more votes per share than a share of regular common stock.

takeover defenses: Strategies for fighting hostile takeovers.

target capital structure: The desired optimal mix of debt and equity financing that most firms attempt to achieve and maintain.

target company: The firm in a merger transaction that the acquiring company is pursuing.

target dividend-payout ratio: A policy under which the firm attempts to pay out a certain percentage of earnings as a stated dollar dividend, which it adjusts toward a target payout as proven earnings increases occur.

target weights: Either book or market value weights based on desired capital structure proportions; used in calculating the weighted average cost of capital.

tax loss carryback/carryforward: A tax benefit that allows corporations experiencing operating losses to carry tax losses back up to 3 years and forward for as many as 15 years. In a merger, the tax loss of one of the firms that can be applied against a limited amount of future income of the merged firm either over 15 years or until the total tax loss has been fully recovered, whichever is shorter.

tax on sale of old asset: Tax that depends upon the relationship between the old asset's sale price, initial purchase price, and *book value*.

technical insolvency: Business failure that occurs when a firm is unable to pay its liabilities as they come due.

temporal method: A method that requires specific assets and liabilities to be translated at historical exchange rates and that foreign-exchange translation gains or losses be reflected in the current year's income.

tender offer: A formal offer to purchase a given number of shares of a firm's stock at a specified price.

term (long-term) loan: A loan made by a financial institution to a business and having an initial maturity of more than one year.

term loan agreement: A formal contract, ranging from a few to a few hundred pages, specifying the conditions under which a financial institution has made a long-term loan.

term structure of interest rates: The relationship between the interest rate and the time to maturity.

terminal cash flow: The after-tax nonoperating cash flow occurring in the final year of a project, usually attributable to liquidation of the project.

time line: A horizontal line on which time zero is at the leftmost end and future periods are shown as you move from left to right; can be used to depict investment cash flows.

time-series analysis: Evaluation of the firm's financial performance over time, utilizing financial ratio analysis.

times-interest-earned ratio: Measures the firm's ability to make contractual interest payments.

total asset turnover: Indicates the efficiency with which the firm uses all its assets to generate sales.

total cost: The sum of the order costs and carrying costs of inventory.

total leverage: The potential use of *fixed costs, both operating and financial,* to magnify the effect of changes in sales on the firm's earnings per share (EPS).

total quality management (TQM): The application of quality principles to all aspects of a company's operations.

total risk: The combination of a security's nondiversifiable and diversifiable risk.

transactions motive: A motive for holding cash or near-cash—to make planned payments for items such as materials and wages.

transfer prices: Prices that subsidiaries charge each other for the goods and services traded between them.

treasurer: The officer responsible for the firm's financial activities, such as financial planning and fund raising, making capital expenditure decisions, managing cash, managing credit activities, and managing the pension fund.

Treasury bills: U.S. Treasury obligations issued weekly on an auction basis, having varying maturities, generally under a year, and virtually no risk.

Treasury notes: U.S. Treasury obligations with initial maturities of between one and ten years, paying interest at a stated rate semi-annually, and having virtually no risk.

treasury stock: The number of shares of outstanding stock that have been repurchased and held by the firm; shown on the firm's balance sheet as a deduction from stockholders' equity.

trust receipt inventory loan: An agreement under which the lender advances 80 to 100 percent of the cost of the borrower's relatively expensive inventory items in exchange for the borrower's promise to immediately repay the loan, with accrued interest, upon the sale of each item.

trustee: A paid individual, corporation, or commercial bank trust department that acts as the third party to a bond indenture to ensure that the issuer does not default on its contractual responsibilities to the bondholders.

two-tier offer: A tender offer in which the terms offered are more attractive to those who tender shares early.

U.S. Treasury bills (T-bills): Short-term IOUs issued by the U.S. Treasury; considered the risk-free asset.

uncorrelated: Describes two series that lack any relationship or interaction and therefore have a *correlation coefficient* close to zero.

underpriced: Stock sold at a price below its current market price.

undersubscribed issue: A security issue whose shares are not immediately sold.

underwriting: The process in which an investment banker buys a security issue from the issuing firm at a lower price than that for which he or she plans to sell it, thereby guaranteeing the issuer a specified amount from the issue and assuming the risk of price changes between the time of purchase and the time of sale.

underwriting syndicate: A group of investment banking firms, each of which will underwrite a portion of a large security issue, thus lessening the risk of loss to any single firm.

unlimited funds: The financial situation in which a firm is able to accept all independent projects that provide an acceptable return.

unlimited liability: The condition imposed by a sole proprietorship (or general partnership) allowing the owner's total wealth to be taken to satisfy creditors.

unsecured loan: A loan that has no assets pledged as collateral.

unsecured, or general, creditors: Creditors who have a general claim against all the firm's assets other than those specifically pledged as collateral.

unsecured short-term financing: Short-term financing obtained without pledging specific assets as collateral.

valuation: The process that links risk and return to determine the worth of an asset.

value dating: A procedure used by non-U.S. banks to delay, often for days or even weeks, the availability of funds deposited with them.

variable growth model: A dividend valuation approach that allows for a change in the dividend growth rate.

vertical merger: A merger in which a firm acquires a supplier or a customer.

voluntary reorganization: A petition filed by a failed firm on its own behalf for reorganizing its structure and paying its creditors.

voluntary settlement: An arrangement between a technically insolvent or bankrupt firm and its creditors enabling it to bypass many of the costs involved in legal bankruptcy proceedings.

warehouse receipt loan: An arrangement in which the lender receives control of the pledged inventory collateral, which is warehoused by a designated agent on the lender's behalf.

warrant premium: The difference between the actual market value and theoretical value of a warrant.

weighted average cost of capital (WACC): Reflects the expected average future cost of funds over the long run; determined by weighting the cost of each specific type of capital by its proportion in the firm's capital structure.

weighted marginal cost of capital (WMCC): The firm's weighted average cost of capital (WACC) associated with its next dollar of total new financing.

white knight: A takeover defense in which the target firm finds an acquirer more to its liking than the initial hostile acquirer and prompts the two to compete to take over the firm.

wire transfers: Telegraphic communications that, via bookkeeping entries, remove funds from the payer's bank and deposit them into the payee's bank, thereby reducing collection float.

work-in-process inventory: All items that are currently in production.

working capital: Current assets, which represent the portion of investment that circulates from one form to another in the ordinary conduct of business.

yield curve: A graph of the *term structure of interest rates* that depicts the relationship between the *yield to maturity* of a security (y-axis) and the time to maturity (x-axis); it shows the pattern of interest rates on securities of equal quality and different maturity.

yield to maturity (YTM): The rate of return investors earn if they buy a bond at a specific price and hold it until maturity. Assumes that issuer makes all scheduled interest and principal payments as promised. Compound annual rate of interest earned on a security purchased on a given day and held to maturity.

zero-balance account: A checking account in which a zero balance is maintained and the firm is required to deposit funds to cover checks drawn on the account only as they are presented for payment.

zero (or low) coupon bonds: Bonds that are issued with no (zero) or a very low coupon (stated interest) rate and are sold at a large discount from par.

zero-growth model: An approach to dividend valuation that assumes a constant, nongrowing dividend stream.